BERNICE SCHACTER

THE NEW MEDICINES

How Drugs Are Created, Approved, Marketed, and Sold

Westport, Connecticut
London

Library of Congress Cataloging-in-Publication Data

Schacter, Bernice Zeldin, 1943–
   The new medicines : how drugs are created, approved, marketed,
and sold / Bernice Schacter.
      p. cm.
   Includes bibliographical references and index.
   ISBN 0–275–98141–X (alk. paper)
   1. Drug development—Popular works. 2. Clinical trials—Popular
works. 3. Pharmaceutical industry—Popular works. 4. Consumer
education. I. Title.
   [DNLM: 1. Drug Industry—organization & administration—United
States. 2. Pharmaceutical Preparations—economics—United States.
3. Clinical Trials—United States. 4. Drug Design—United States.
5. Drugs, Investigational—economics—United States. 6. Legislation,
Drug—United States.   QV 736 S291n 2006]
   RM301.25.S34 2006
   615'.19—dc22   2005019187

British Library Cataloguing in Publication Data is available.

Library of Congress Catalog Card Number: 2005019187
ISBN: 0–275–98141–X

First published in 2006

Praeger Publishers, 88 Post Road West, Westport, CT 06881
An imprint of Greenwood Publishing Group, Inc.
www.praeger.com

Printed in the United States of America

The paper used in this book complies with the
Permanent Paper Standard issued by the National
Information Standards Organization (Z39.48–1984).

10   9   8   7   6   5   4   3   2   1

Contents

Preface

Do we have the pharmaceutical industry we need and want? Perhaps not, but it is the industry that our innovative and entrepreneurial society and responsive regulatory system has given us. And knowing how it works may allow us to make better decisions as patients, caregivers, and citizens.

The seed for this book began in 1984 when, with little forethought, I landed in the midst of an industrial effort, the fruits of which had benefited me but the workings of which were pretty much unknown to me. After serving in an academic position where I did research, taught immunology to medical students, and ran a laboratory that matched people for organ transplants, I moved to a research scientist position at what was then Bristol-Myers, where I was quickly exposed to the steps of drug discovery and development. With no wall separating the research and development work, department and division review meetings provided ample opportunities for a glimpse of what it took to morph an idea into a drug in a bottle. It did not hurt that my spouse was immersed in the same process on the clinical side. "Phases," "the Agency," and "project status" became my jargon, though my nominal assignment was as a lab jock and I worked hard at it. It also did not hurt that through my academic experience and by propinquity (i.e., marriage) I had learned a little bit of clinical medicine (always a dangerous thing) or that Bristol licensed a putative anticancer drug that I thought demanded my championing for selection for development as a drug for organ transplantation patients. I am nosy and always want to know about everyone's job. People like talking about what they do, and even if they grouse, a little sympathy and a few questions will get them explaining over a cup of coffee.

After seven years, the maelstrom of a corporate merger (with Squibb), two years leading R&D at a biotech start-up, and a chronic illness that precluded the career fast track, I was back in the classroom, eventually teaching science to nonscientists. What better pedagogic platform than "How did that pill get into the bottle?" More homework (lots of homework—bless the Internet and online libraries), and I had a course. Then a book seemed a nifty way to pull the curtain back for a larger audience and expose the wizard that puts the pills in the bottle. Current posturing by the usual and always righteous suspects notwithstanding, the wizard is neither inherently evil nor unusually fumbling. The wizard is no saint either. Getting the pill in the bottle and into your medicine cabinet and mine takes the dedicated, clear-eyed efforts of scores of smart, generally conscientious, and often very ambitious people who are as fallible as any such group. I think the story of their efforts and fumbles is worth knowing because as citizens, investors, and, yes, as patients, we have to make judgments about how well they do their jobs of moving the drug bit by bit along its proscribed and treacherous path.

I must acknowledge the help and counsel of many (now former) Bristol-Myers people for my education, particularly, but in no particular order: Stephen Carter, Sal Forenza, Al Crosswell, Sue Hubencz, Cheryl Anderson, Robert Wittes, Terry Doyle, Anna Casazza, Bill Rose, John Shurig, George Spitalny, Joe Brown, Alice Leung, and so many others who either patiently explained some nuance of their part of the process or, in the case of my supervisors, expected that I would learn all this to better do my job.

In developing this book, I shamelessly called on the expertise and patience of Kevin O'Neill, Jerry Merritt, Frank Bamborolla, Susan Anderson, Donna Francher, and Lottie Wang. I am also deeply indebted to John Talley, Stuart Scheindlin, and Irit Pinchasi not only for their patience but also for their willingness and ability to recall events in the distant past. Thanks, too, to Kevin Downing for his wise counsel and skillful editing.

Beyond all, I must give profound thanks to my husband, Lee P. Schacter, Ph.D., M.D., for creating the opportunity to enter this world and for sharing his education in drug development with me. We "talk" science endlessly, as our patient daughters can attest, and share the professional challenges we face. Lee is smart, wise, knowledgeable, and ethical, a combination hard to match. He taught me that the ethics of medical research not only allow rigorous science but also require it. For that I give my humble and loving thanks. As for his patience when I am writing, I can only ask, again, "It wasn't so bad, was it?" So, too, I ask Beth and Sara.

Abbreviations and Acronyms

AD   Alzheimer's Disease
ADME   Absorption, Distribution, Metabolism, and Elimination
ALL   Acute Lymphoblastic Leukemia
AMP   Average Manufacturer Price
ANDA   Abbreviated New Drug Application
AUC   Area Under the Curve
AWA   Animal Welfare Act
BLA   Biological License Application
CBER   Center for Biologics Evaluation and Research
CDC   Centers for Disease Control
CDER   Center for Drug Evaluation and Research
CFR   Code of Federal Regulations
CGCP   Current Good Clinical Practice
CGLP   Current Good Laboratory Practice
CGMP   Current Good Manufacturing Practice
COX   Cyclooxygenase
CRF   Case Report Form

CRM	Customer Relationship Marketing
CRO	Contract Research Organization
DSMB	Data Safety Monitoring Board
DTC	Direct to Consumer
EAE	Experimental Allergic Encephalomyelitis
FACA	Federal Advisory Committee Act
FDA	Food and Drug Administration
FDCA	Food, Drug, and Cosmetics Act
FTC	Federal Trade Commission
GAO	General Accounting (now Accountability) Office
GI	Gastrointestinal
HED	Human Equivalent Dose
HEW	U.S. Department of Health, Education, and Welfare (now HHS)
HHS	U.S. Department of Health and Human Services
HIPAA	Health Insurance Portability and Accountability Act
HMO	Health Maintenance Organization
IACUC	Institutional Animal Care and Use Committee
IB	Investigator's Brochure
ICH	International Conference on Harmonization
IND	Investigational New Drug (application)
IOM	Institute of Medicine (of the National Academy of Sciences)
IRB	Investigation Review Board
ISS	Integrated Safety Summary
ITT	Intent to Treat
MBP	Myelin Basic Protein
MS	Multiple Sclerosis
NAS	National Academy of Sciences
NCI	National Cancer Institute
NDA	New Drug Application

NIH	National Institutes of Health
NMSS	National Multiple Sclerosis Society
NOAEL	No Observable Adverse Effect Level
NRC	National Research Council (of the National Academy of Sciences)
NSAID	Nonsteroidal Anti-inflammatory Drug
OGE	Office of Governmental Ethics (U.S.)
OIG	Office of the Inspector General (U.S.)
OTA	Office of Technology Assessment
OTC	Over the Counter (not needing a prescription)
PBM	Pharmacy Benefits Manager
PD	Pharmacodynamics
PDMA	Prescription Drug Marketing Act
PDUFA	Prescription Drug Users Fee Act
PFDA	Pure Food and Drug Act
PGE2	Prostaglandin-E2
PhRMA	Pharmaceutical Research and Marketing Association
PK	Pharmacokinetics
PML	Progressive Multifocal Leukoencephalopathy
RTF	Refuse to File
SGE	Special Government Employees (U.S.)
sNDA	Supplemental New Drug Application
SOP	Standard Operating Procedure
USC	United States Code
USDA	U.S. Department of Agriculture

1

The Path from Bench to Bedside

To have his path made clear for him is the aspiration of every human being in our beclouded and tempestuous existence.

Joseph Conrad, *The Mirror of the Sea*, chap. 27 (1906)

A SORE KNEE

Imagine the prototypical "average" American. Let us call him Roger. Roger is a man in his fifties, relatively healthy, and with no serious medical problems. He keeps in shape by jogging—every morning, he ties on his running shoes for a three-mile run around the neighborhood. He has done this for years and has never had a physical problem that could not be solved with a couple of aspirin.

But now that was changing because his left knee hurt. It really was swollen and was stiff for about half an hour after he got up. When he bent it, it made that old man creaking sound. When he ran, and actually even when he walked, it just hurt.

Roger tried the usual stuff from the drug store—aspirin, Tylenol, Advil. Although they helped ease the pain, he remembered reading that too much aspirin or Advil could hurt his stomach. He remembered that his aunt developed an ulcer after taking a lot of aspirin for years for her arthritis. Anyway, none of these remedies helped much.

Roger made an appointment with his doctor, and after a thorough examination, including blood tests and X-rays, he received the diagnosis—he had arthritis, a type called osteoarthritis. For some reason, the cartilage, the gristle, in his knee had worn away, and the bones in the knee were rubbing each other when he walked or ran, or just when he bent the knee. The cracking sound—which his doctor said was sometimes called joint mice by the older docs, was caused by bone rubbing bone.

Roger's doctor recommended a new drug, Celebrex. It worked like aspirin and Advil, but with less chance of stomach problems. Roger had seen the ads on TV; they made it seem like magic, but there were all those warnings at the end. It was not the expense—he had prescription coverage with his health insurance—he was just unsure about all those warnings. For one thing, his doctor had said he would have to have regular blood tests to make sure that the pill was not harming his liver or kidneys. Also, Merck, the company that made a similar drug, Vioxx, had just stopped selling their product. All of this made Roger feel very old. It was one thing to take some aspirin for an ache or pain but another thing to take a prescription drug for arthritis.

How did the doctors figure out that this was the right drug for what Roger had? How did his doctor learn about Celebrex? Was a salesman from the drug company selling him a bill of goods? If the company could make those ads with the women doing t'ai-chi, maybe the salesmen could dazzle the doctors. If Roger only knew how the system worked, maybe he could trust his doctor's suggestion.

This is a situation that most of us find ourselves in at one point or another. How drugs are discovered, tested, and approved is a process that is obscure to most of us. We read the news articles that often seem like press releases, claiming a new, improved treatment for disease X. The article often concludes with a caution that it will be two, three, or sometimes five years before testing is complete and the drug is approved. What takes so long? What is the process for the testing and approval of a drug? What are the rules for the testing, and who makes those rules? Who decides whether a drug should be approved, and how do they reach their decision? Who protects us from useless or unsafe drugs? This book will answer these questions and will help the reader understand how new medicines go from the laboratory bench to our medicine cabinets.

THE STORY OF THE BREAKTHROUGH: HOW DID THEY FIND THE TREATMENT FOR X?

The story about today's breakthrough probably started a decade earlier, in a chemistry or biology laboratory where a scientist had an idea about a particular normal or disease process in the body, an idea that suggested that a drug that could interrupt one process or trigger another process might influence a disease. The resulting research—testing the idea, that hypothesis about how the process worked—took years of experiments with human or animal cells and cell components and then with whole animals that had something similar to the human disease. Only then did the search begin for a chemical compound that worked as the researchers hoped, both in the test tubes and in the animal. That might lead to a eureka moment: "We have a drug!"

In fact, that eureka moment in the lab was only the beginning. Now business and industry take over because the process of moving the chemical from eureka to a bottle at the drug store is highly regulated, complex, expensive, and risky. Only a small fraction of such discoveries by scientists move from the research stage to the formal process of laboratory and clinical testing that must be done to gain approval from the government to sell the drug. And once the drug enters human testing, the risk of failure remains high. The probability of a successful market launch for a new drug that begins human testing is estimated to range from 10 to 25 percent. The success rate depends on many factors: the time period analyzed, the type of drug (small chemical or large protein), and the depth of understanding of the process underlying the disease that the drug is intended to treat (CDER 1999a; DiMasi 2001; Grabowski and Vernon 1990; OTA 1993; U.S. Congress 1992). One industry executive compared the risk to "wildcat" oil exploration in Texas (Hilts 1992).

The steps to approval, which are well-described in government regulations and guidelines, are diagrammed in Figure 1.1.

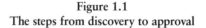

Figure 1.1
The steps from discovery to approval

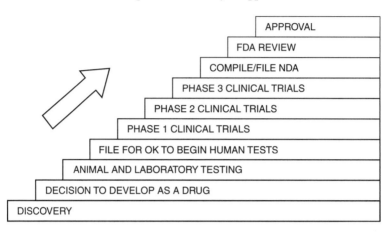

Although the medical market, the physicians, insurance companies, and, yes, the patients are the final determinants of the success of a new drug, the initial gatekeepers are the government regulatory agencies. The Food and Drug Administration (FDA) in the United States and its counterpart regulatory agencies in Japan (Ministry of Health and Welfare), Canada (the Health Protection Branch), and in Great Britain and other members of the European Community regulate the sale of pharmaceuticals and biopharmaceuticals for human use based on scientifically sound evidence of safety and effectiveness. In the United States, the FDA regulates new therapeutic and diagnostic agents through its centers. The Center for Drug Evaluation and Research (CDER) handles evaluation of both drugs (small, naturally found or synthetic compounds with defined structure and physical and chemical properties) and, as the result of a 2003 reorganization, biologics such as laboratory-generated proteins and antibodies. The Center for Biologics Evaluation and Research (CBER) is responsible for serums and vaccines. The regulations controlling the production of pharmaceuticals and biopharmaceuticals for human use provide safeguards to assure the identity, strength, quality, and purity of the agent.

The preclinical studies for initial clinical testing must meet strict requirements in terms of performance, outcome, and documentation. The clinical trials must be rigorously designed and carried out with a goal of providing a prudent, stepwise, and statistically sound demonstration of safety and effectiveness for the proposed use. This book focuses on the development and registration of agents for human use in the United States. Foreign regulatory agencies address many of the same concerns as the FDA, though there are differences, particularly in the balance of concern for scientifically sound evidence of efficacy as well as safety. There is an ongoing, concerted international effort to harmonize the regulations.

SHOULD WE BUILD IT? THE DECISION TO DEVELOP A DRUG

The process starts with the discovery of a potential new drug in research laboratories within a company, a university, or a government lab. But the decision by a pharmaceutical or biotechnology company, the sponsor, to move a compound from research to the status of a project—to commit to the next stage, preclinical development—is generally not lightly made. The company considers many issues, including the degree of difficulty in manufacturing the product, the likelihood of obtaining regulatory approval for the product, and the potential for its successful marketing. The company must be able to put together a workable plan for the human trials needed to demonstrate that the product has a medical use and that it is safe for that use.

Moving from research to project status requires a great deal of laboratory work, the establishment of protection of the intellectual property through patent filings, and the rigorous assessment of the potential clinical use of the agent. The financial burden and benefit to the healthcare system of new treatments are of growing concern in today's era of cost containment, and analyses of the economic costs and benefits of the product are critical and will be used to develop clinical strategies and the design of clinical trials. These issues are considered early and reconsidered often.

The likely target disease or condition and the clinical need for new therapies for the specific condition and stage of the condition must be explored. How many people have the disease? Are their clinical needs unmet by currently available treatments? The patient population(s) thought to be candidates for the agent must be described, quantified, and located in the health care system. Issues such as whether the patients will be treated as inpatients or outpatients, at home or in a physician's office, and whether the condition is life-threatening must be addressed. These data form a foundation of assumptions that are repeatedly questioned and challenged as more experience with the potential agent in animal models, animal safety testing, and clinical trials is gained and as the field changes with the passage of time.

A number of things must be accomplished before human trials can begin. A way must be found to manufacture the drug in sufficient amounts by procedures that will meet FDA requirements. Laboratory tests and animal studies must be done to develop an understanding of how the drug works—what it does to bring about its effects. Animal safety studies must be done and a plan developed for how the human trials will be done.

Preclinical animal safety studies provide evidence that it is, in the language of the regulations, "Reasonably safe to conduct the proposed clinical trials." The specific animal studies that will satisfy the concerns of the FDA for new drugs generally include acute and subacute toxicity studies. In acute toxicity studies, the drug is administered once in increasing doses from a no-effect dose to doses causing major life-threatening effects. Subacute toxicity studies describe harmful effects that appear after longer exposure and determine whether and how the drug-induced toxicities progress and resolve. The duration of treatment and the dosing for subacute studies will depend on the planned human dosing and schedule, and usually require at least twice the planned human duration of exposure. Other animal safety and laboratory studies, including chronic toxicity studies, tests of the potential to cause tumors, effects on reproduction (including on fetuses), and potential to cause damage or even changes in the genetic material of cells and tissues often begin later in the process of development, after the initiation of Phase 2 clinical trials (Mathieu 1994).

Because of species specificity and the potential to lead to allergic reactions, biologics pose particular challenges in establishing that it is reasonably safe to begin human trials, and the animal safety tests needed for each biologic material are addressed on a case-by-case basis.

BEFORE CLINICAL TRIALS MAY BEGIN

Clinical (human) trials are used to establish the safety and efficacy of the new agent. Before clinical trials on a new drug may begin (i.e., before the drug or biological may be shipped across state lines), the sponsor must file an IND (Investigational New Drug application) with the FDA. In the IND, the sponsor of the clinical trial provides the FDA with substantial evidence that the drug has been manufactured, can be provided to the clinical investigators with consistent purity and potency, and has been sufficiently tested in animals to anticipate its potential significant toxicities in humans and to understand how the drug is metabolized and excreted. In addition, the IND describes the proposed clinical use for the drug and provides detailed outlines of the initial clinical trials. The FDA has 30 days from the filing date to review an IND. If the FDA finds significant deficiencies in any of the information provided, the drug is placed on "clinical hold" until these concerns are resolved.

The trials generally proceed in three phases, which are designed to establish the safety, dose, and utility of the agent, scientifically and with minimal risk to subjects (Mathieu 1994). Each phase has specific goals and features (Figure 1.2).

By law, investigational review board(s) (IRB) of the hospitals or research institutions where the studies will be performed must review and approve proposed studies at each phase. These boards are independent of the sponsor and are responsible for assuring that each study presents no unacceptable risks to the subjects and that the subjects are informed and willing to participate, as documented in writing.

Figure 1.2
Phases of clinical development

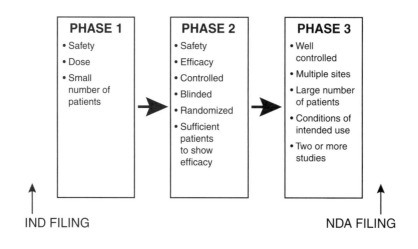

PHASE 1
- Safety
- Dose
- Small number of patients

PHASE 2
- Safety
- Efficacy
- Controlled
- Blinded
- Randomized
- Sufficient patients to show efficacy

PHASE 3
- Well controlled
- Multiple sites
- Large number of patients
- Conditions of intended use
- Two or more studies

IND FILING   NDA FILING

PRODUCTION

The production of a potential new therapeutic product must be achieved by methods that provide assurance of satisfactory purity and consistency of the drug itself and the drug product, the mixture of the drug, and all the additives in the pill, cream, or solution that is provided to the patient. This must be done at the highest level of performance and documentation in order to meet FDA requirements and be cost-effective. The regulations established for the performance and documentation of production of a drug or biologic approved for human use are termed good manufacturing practices (GMP) (Mathieu 1994). The material used in animal safety testing and in clinical trials done to obtain approval must meet certain GMP requirements.

PHASE 1 CLINICAL TRIALS

Phase 1 trials establish whether the drug has any negative effects on the subjects, how quickly the drug enters and leaves the human body, and what effects it has on the target cells or proteins. Depending on the drug, these trials may enroll healthy control subjects or subjects with the condition for which the drug may eventually be used. Phase 1 studies usually involve a small number of subjects, generally 20–100. The goal is to define a safe dose and a schedule for dosing that has the desired effect on the drug's target cell or protein.

PHASE 2 CLINICAL TRIALS

On the basis of a careful review of the data from the Phase 1 trials, the sponsor will begin Phase 2 trials in subjects who have the condition for which the drug is intended. Phase 2 trials establish evidence of the safety and effectiveness of the drug for the intended use. The dose (the amount of drug), the route (how it will be given, for example by mouth or by injection), and the schedule (how many times and how frequently it will be given) may be varied to achieve adequate blood levels of the drug for an appropriate period of time to obtain the desired effects on the target protein, cell, or tissue. The Phase 2 studies are generally randomized, double-blind, controlled studies. The term *controlled* means that some of the patients will receive the drug under study and some will receive either a placebo (an inactive material, such as a salt solution or a sugar pill) or the currently accepted treatment for the condition. *Randomization* means that the assignment of each patient to the experimental or control arms of the study will be entirely by chance and that in all important respects (such as age, sex, disease state, and other factors that could affect the course of the illness and the effects of the drug) the groups are matched. *Double blinding* means that neither patients nor the physicians, nurses, or other health care providers involved in the study know the arm in which any patient is enrolled. Blinding assures that assessments of both safety and efficacy will be made without any conscious or unconscious bias on the part of the patient or the health care providers. The purpose of Phase 2 studies is to test in a scientifically rigorous manner whether the experimental drug has beneficial activity and safety in patients for whom the drug is intended. Ideally, Phase 2 studies are statistically robust tests of the null hypothesis, the hypothesis that the new agent is *not* more effective than the control. The goal is to exclude the null hypothesis. Phase 2 trials may involve up to several hundred patients and may take several months to two years to complete. Phase 2 trials essentially establish that the agent has potential as a useful drug and indicate under what settings and for which conditions it may be demonstrated to have such utility in Phase 3.

PHASE 3 CLINICAL TRIALS

Phase 3 trials are more extensive, randomized, well-controlled, double-blinded evaluations of the long-term safety and effectiveness in a larger number of subjects who are representative of the patients seen in ordinary medical practice. The dose and schedule will be further defined. Less frequent and later side effects may be discovered. A realistic assessment of the benefits versus the risks of taking the drug is obtained for a large population of patients for whom the drug is intended. Phase 3 studies may enroll several thousand subjects and take one to four years to complete, in part because of the need for longer-term follow-up of patients. Again the ideal is a rigorous test of the null hypothesis. Overall, the clinical trials may take two to ten years, with an average of around five years. The time will depend on the nature of the drug, the condition for which it is intended, and its intended duration of use. During the clinical trials, the longer-term animal safety studies will also be performed.

FILING FOR APPROVAL AND REVIEW BY THE FDA

At the conclusion of the clinical trials, including the follow-up period, the sponsor assembles all the relevant information about the drug, including the manufacturing, animal safety testing, and the results of all clinical trials, into a new drug application (NDA). The NDA can run 100,000 pages or more. The application is thoroughly reviewed by FDA scientists and physicians. The FDA reviewers, including very experienced clinical biostatisticians, will carefully review the power of the studies and the validity of the conclusions. An advisory committee of outside experts will also review many NDAs. At the advisory committee meetings, the sponsor's physicians and scientists present the essential information about the clinical trials and answer questions from committee members, who after extensive discussions vote on whether to recommend approval of the application for specific conditions supported by the data. The FDA is not required to accept the recommendation of its advisory committees, but it generally does approve as recommended. The review of the application by the experts at the FDA and their advisors can take from a few months to seven years, depending on the complexity of the program and whether an NDA has to be resubmitted after an initial review and rejection. The FDA reported that in 2004 the median time from NDA filing to approval was 12.9 months (FDA 2005a).

IT IS AN UNCERTAIN PROCESS

Clearly, the development of a new drug, whether conventional or biologic, is lengthy, expensive, and risky. The FDA indicates that approximately 20 percent of drugs entering Phase 1 clinical trials in the United States will ultimately be approved for use in humans. The success rate for biologics, proteins and other drugs produced using genetic engineering methods may be a bit better, but the biotechnology industry is relatively new, and early biologics exploited some very promising medical niches. Approval by the FDA will allow launch and marketing, and the sponsor must have production, marketing, and sales capacity in place to successfully launch a new drug.

The process of developing a new drug or biologic can take more than 12 years and is reported to cost over $800 million (U.S.) (Figure 1.3) (DiMasi, Hansen, and Grabowski 2003). The regulations covering the care and scientific rigor of the production and testing of conventional new drugs and biologics for human use have been developed to protect patients from unsafe and ineffective therapies. The FDA has been criticized both for being too slow and cautious in the evaluation of new drugs and biologics and for approving drugs too rapidly, without enough evidence of safety or efficacy. Suggestions for streamlining the process and for making it more rigorous are popular topics for discussion in the pharmaceutical and biotechnology industries. The media, in response to the latest report of toxicity uncovered only after approval of a new drug or of the unavailability of a drug in the United States despite its approval elsewhere, will argue both ways.

Figure 1.3
Timeline for development of a new drug for humans

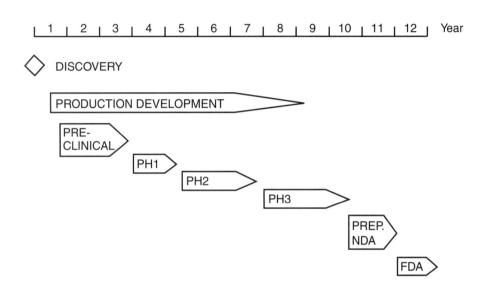

Stories of an unsafe cholesterol-lowering drug and arthritis pills that may increase the risk of heart attacks and stroke have fueled challenges by the media and politicians about how the safety of a drug is considered. The ideal of rapid approval of perfectly safe and effective drugs remains a dream. Government, industry, and the health care system strive to come as near as possible to the ideal. As with all human endeavors, this is an imperfect process, and now more than ever it is one that commands our attention.

Roger is right to have questions about the medications that his doctor is prescribing for him; as we will see throughout this book, the procedures for getting a drug out of the laboratory and into his medicine cabinet are painstaking and complex. They are also, by and large, hidden from the consumers, who are left wondering whether they are getting safe and effective drugs. The rest of this book will attempt to provide some insight into these issues.

2

How Did the FDA Get to Be in Charge?
The History of Regulation of Human Drugs

The U.S. government regulates the testing, manufacture, approval for sale, and marketing of pharmaceuticals under the Federal Food, Drug, and Cosmetic Act, as amended by the FDA Modernization Act of 1997. Congress has determined that the Food and Drug Administration's roles in regulating prescription drugs are to protect public health by approving prescription drugs that are safe and effective (CDER 1999a).

If you have ever had a prescription filled with the bottle as it came from the drug company, you may have found tucked inside the box or stuck on the lid a neatly folded document filled with paragraph after paragraph of fine print—detailing the active and inactive ingredients of the prescription, the results of studies on the drug, who should or should not take the drug, and what disquieting things might happen if the drug is taken. That document of nearly illegible jargon is called the package insert, or the prescribing information. Current regulations require that a drug must be safe and effective for the use printed on the package insert and that the package insert provides what the Food and Drug Administration has concluded must be easily available to physicians prescribing the drug. The manufacturer must provide such information and is open to prosecution if the drug is adulterated or misbranded, not only in terms of contents but also in terms of the description of approved uses and risks of the drug.

How did the FDA come to be the gatekeeper for the medicines, with broad powers to allow marketing or forbid it and to approve what is said to physicians and to the public about the approved drugs? The history of the FDA, and its ongoing, often adversarial role in regulating not only the drugs we take but the food we eat and the cosmetics we put on our faces, is a study of the tension between industry and regulators in a free market economy (Janssen 1981a, b, c). The story has been very well told by Philip Hilts in *Protecting America's Health* (Hilts 2003) and will be briefly recounted here.

PATENT MEDICINES: CAN YOU TRUST THE SPIEL?

The initial rules on drugs in the United States and Canada, developed before the U.S. Civil War, focused on the issues of adulteration and misbranding, consumer protection issues that arose from accidental or purposeful inclusion of toxic materials in a drug being sold. Before advances in chemistry spawned the pharmaceutical industry, extracts of herbs, plants, and animal materials, mixed with alcohol and other chemicals, were sold at stores and pharmacies. In the nineteenth century, many traditional remedies and newly isolated chemicals were sold as patent medicines, an intentionally misleading term. Many were not patent-protected but were simply secret mixtures and extracts. Initially sold door to door and by vendors in medicine shows, they eventually were also sold in pharmacies, at great profit to the pharmacists. Patent medicines were not regulated with regard to content or the healing claims made by their purveyors. The claims were great—cures for all the ills of humankind—and the prices were low. The industry was large and successful, slowed little by the ire of organizations of orthodox physicians (CDER 2003a).

During the period following the Civil War, physician organizations, such as the American Medical Association, were so embroiled in debates on how to improve the quality and stature of physicians that they paid little attention to the issue of nostrums and alcohol-laced cure-all elixirs being sold to the public (Starr 1982). The German pharmaceutical industry was also developing, shipping effective fever treatments and pain medicines to be prescribed by physicians (Landau et al. 1999). Physicians in the United States had to struggle to learn which medicines were safe and effective and which were useless and perhaps dangerous. Pharmacists were torn between concern that the patent medicines they might sell were useless or even dangerous and the reality that significant profit could be made from their sale. Debates within professional pharmacists' organizations had little effect.

THE LONG FIGHT TO GET SOME TRUTH AND PROTECTION: THE FOOD AND DRUG ACT OF 1906

Patent medicine manufacturers advertised widely and developed the strategy of including a clause in their advertising contract with newspapers and magazines that voided the contract if the state where the paper was published passed a law prohibiting or restricting the sale of patent medicines. This clause, termed the Red Clause, tended to silence the zeal of editors who might wish to urge legislation to restrict the fraudulent claims of the patent medicines; it might hurt their paper's advertising revenue. Enraged at the Red Clause, Norman Hapgood, the editor of *Colliers* magazine, commissioned Samuel Hopkins Adams to write a series of articles as an exposé of the patent medicine business. Adams's articles in 1905 and 1906, called "The Great American Fraud," were a thorough and detailed account of the industry, its tricks and advertising strategies, and the alcohol- and morphine-laced concoctions hawked to an anxious, ill-informed, and gullible populace. He described not only the business of buying and selling testimonials of purportedly satisfied customers but also the solution of dilute acid sold as an antiseptic and the throat and teething syrups laced with cocaine and opium. State and federal governments were drawn into the campaign against patent medicines by the public's response to the series in *Colliers*, but attempts in state legislatures and the U.S. Congress to pass bills directed at assuring the purity of drugs struggled against strong lobbying by the Proprietary Association, the organization of the patent medicine companies (Swann 1998).

Harvey Washington Wiley, who became chief chemist of the Agriculture Department in 1883, spent decades documenting how vulnerable America's food was to adulteration. He became determined to have a pure food and drug bill enacted, and worked relentlessly, through seemingly endless political battles, until in 1906, possibly spurred by the *Colliers* exposé, Congress passed, and President Theodore Roosevelt signed, the Pure Food and Drug Act (PFDA) (Swann 1998; Young 1981).

The Pure Food and Drug Act prohibited the interstate commerce of misbranded and adulterated food, drink, and drugs. Violation of the provisions of the Act were deemed a misdemeanor, conviction bringing a fine of no more than $500 and a year's imprisonment. Subsequent convictions brought a $1,000 fine and up to a year's imprisonment. The reference for adulteration was the test for the drug as published in the U.S. Pharmacopeia's National Formulary. The U.S. Pharmacopeia, established in 1820 by a group of physicians, publishes the National Formulary, a list of standard drugs and associated inactive substances used as vehicles or solvents for the drug. The formulary also describes the purity and strength of the drug, thus providing standards against which marketed drugs can be compared. The U.S. Pharmacopeia (USP), a nongovernmental, volunteer organization, was established as the legal reference in the 1848 Drug Importation Act, passed to empower the U.S. Customs Service to prevent the entry of adulterated drugs into the United States (U.S. Pharmacopeia 2003).

IT FAILED TO DO THE JOB; DISASTER DRIVES CHANGE

The limits of the PFDA were apparent when in 1911 the Supreme Court ruled, in *United States. v. Johnson*, that a false therapeutic claim, in this case a claim for a cure for cancer, was not prohibited by the Act. In 1912, Congress passed an amendment to the Act, the Sherley Amendment, that prohibited only fraudulent therapeutic claims, false claims *intended* to defraud. The Sherley Amendment had little impact because intent to defraud is difficult to prove (Swann 1998).

The next major change to the federal regulation of drugs came in 1938 with the passage of the federal Food, Drug, and Cosmetic Act (FDCA), triggered by the death of over 70 people by poisoning from diethylene glycol (familiar as antifreeze), used by a manufacturer as a solvent for sulfanilamide, an antibiotic, without determining whether the solvent was safe. Under the FDCA, the FDA had to be notified before a new drug could be shipped across state lines for sale, which would be permitted if no objections were raised within 180 days. New drugs had to be shown to be safe before marketing. The Act also eliminated the stipulation that false claims were only illegal if intent to mislead was proven. Drug packages had to carry both instructions for use and warnings of dangers.

In 1951, the Durham-Humphrey amendment to the FDCA was passed, based on the idea that certain drugs could only be safely given under medical supervision and thus could only be sold with a prescription from a physician. This rule applied specifically to drugs that caused sleepiness, drugs that were habit-forming, drugs that were toxic or harmful, or drugs that had not been shown to be safe for self-medication. This amendment was supported by organizations of pharmacists because it clarified what pharmacists could legally sell "over the counter." The terms of the amendment were confirmed by a U.S. Supreme Court decision upholding the conviction of a pharmacist named Sullivan for dispensing 12 sulfathiazone pills in an unlabeled box, the pills having been purchased out of state and thus covered by the federal FDCA (Smith 2002; *U.S. v. Sullivan* 1948).

BIG BUSINESS GETS NOTICED, BUT IT TAKES ANOTHER DISASTER TO MAKE NEEDED CHANGES

After World War II, the pharmaceutical industry experienced significant growth and many new drugs and new classes of drugs were developed, particularly antibiotics, hormones, and drugs for pain. Some of the practices of the growing industry raised consumer concerns and caught the attention of the Senate Subcommittee on Antitrust and Monopoly. The high prices charged by the pharmaceutical companies, particularly for new drugs, were the focus of Subcommittee hearings in the 1950s (Harris 1964). The chairman of the Subcommittee, Estes Kefauver, drafted legislation to modify the marketing practices of the pharmaceutical companies, but the legislation was stalled by political wrangling over Kefauver's sweeping proposals for changes in the FDCA that would lower prices for medicines by increasing competition as well as expand the authority of the government to protect the public from dangerous medicines. One particularly

contentious proposal was the shortening of the patent-protected monopoly for drugs by requiring the patent holder to provide licenses to
others to manufacture and sell the drug after three years of exclusivity. To address the practices of the drug companies that led to higher
prices and stifled competition, Kefauver wanted to amend the Sherman Antitrust Act to make it unlawful for drug companies to pri-
vately settle patent disputes. To limit the growing practice of marketing
new, higher-priced drugs with little improved benefit over existing
drugs, he proposed that no modified or combined drugs should be
approved for sale unless they were proven to provide a significant improvement. The names of new drugs were also addressed. The pharmaceutical companies had developed a practice of naming their drugs
with noninformative and confusing names that did not reflect the
chemical nature of the drug. The flurry of trade names meant that
physicians had to rely on the drug salesman's pitch to learn of the
benefits of each new drug. Kefauver proposed that the secretary of
the Department of Health, Education, and Welfare (HEW) assign
each drug a generic name so that different companies' versions of
the drug would carry the same name and information provided to
the physician would carry both the generic name and the trade
name. Drug manufacturers would be licensed, would be required to
provide information on how they assured the purity of the drugs, and
would have to allow inspections by agents from HEW. Kefauver also
proposed that HEW publish a list of harmful drugs and that advertisements include information on side effects as well as claims for the
benefits of the drug. These changes were well beyond what the Senate
was prepared to enact. Industry lobbying was predictably intense. But
fate created a crisis that provided the public outrage and political mo-
mentum for passage of a set of amendments that included some but
not all of the reforms Kefauver sought (Harris 1964).

   In 1962, evidence was published that severe birth defects in thousands of infants were caused by a sleeping pill, thalidomide, that had
been approved in some European countries but not in the United
States. Only the persistent efforts of an FDA medical reviewer, Dr.
Francis O. Kelsey, had prevented the approval of thalidomide in the
United States. Unfortunately, the company developing thalidomide
had been sending samples of the drug to U.S. physicians to give to

their patients in order to accumulate more information for marketing. This unethical, unapproved, unregulated human drug trial, without the knowledge and consent of the patients, resulted in the birth in the United States of hundreds of children with devastating birth defects (Harris 1964). This tragic reminder of the need for diligent regulation of pharmaceutical agents was sufficient to drive passage of the Kefauver-Harris Drug Amendments, which required evidence of both efficacy and safety before a drug could be marketed, the collection and submission to the FDA of adverse reaction reports by drug firms, inclusion of complete information in drug advertising in medical journals—the risks as well as the benefits of the drugs, the withdrawal from the U.S. market of drugs lacking evidence of safety and/or effectiveness, and the review and revision by the HEW of the official (generic) names of drugs (Hilts 2003; Swann 1998). Most significantly, the process changed from premarketing notification to premarketing approval, based on a balanced assessment of efficacy and safety. Although other modifications to the FDCA have been made, this set of amendments provides the fundamental enabling law for the FDA to regulate the approval of pharmaceuticals (Harris 1964).

CHARGING THE REGULATED TO GET TIMELY REVIEWS

The responsibility of the FDA to provide for prompt approval of safe and effective drugs was codified in the 1992 FDA Modernization Act, which included The Prescription Drug Users Fee Act, which established a fee structure for review of applications to test, manufacture, and market pharmaceuticals. The fees provide funds to allow the FDA to employ more scientific and medical reviewers and thus complete reviews more quickly. The Prescription Drug Users Fee Act was reauthorized in the 1997 FDA Modernization Act, which also addressed the issue of advertising approved drugs for nonapproved uses (Food and Drug Administration Modernization Act of 1997). The Act was reauthorized in 2002 in the Public Health Security and Bioterrorism Preparedness and Response Act (Public Health Security and Bioterrorism Preparedness and Response Act 2002).

The rules governing the testing, approval, manufacturing, and marketing of drugs are provided at three levels: the Food, Drug, and Cosmetic Act as amended by the FDA Modernization Act of 1997; regulations published in the Federal Register; and guidelines available from the FDA. The regulations, found in Title 21, Chapter 9 of the Code of Federal Regulations (CFR), provide a detailed description of requirements for all aspects of the manufacture, testing, approval, and marketing of pharmaceutical drugs. The guidelines, covering a broad range of topics from advertising to user fees, do not have the force of law but are provided to the sponsor, the company seeking approval, to share the FDA's current thinking on a subject. Of course, court decisions also establish the limits of statutory authority of the Food and Drug Administration.

Just as scandals, controversies, and crises drove the evolution of the rules that control the approval of new drugs in the past, each year, or sometimes each month, seems to bring an alarm that the FDA is not doing enough to protect us from unsafe or useless drugs, as well as attacks that the FDA's rules stifle the development of new drugs and keep useful drugs from us. It almost seems that the FDA can please no one, being criticized as either too lax or too strict. They must perform a careful balancing act, buffeted by politics, which is the lot of any civil agency (Carey 2003). The alarms that were first raised in 2002 about certain types of pain relievers provide a timely example of what can happen. We will address this topic as our story unfolds.

3

The Eureka Moment:
How New Medicines Are Discovered

Consider Lynne, a recent Ph.D. with training in biochemistry who has been hired by a pharmaceutical company to run a drug discovery laboratory. She has left university life and the specter of never-ending rounds of grant-writing to supervise a group of technicians testing vast collections of compounds to find those that block the production of prostaglandins, a major player in the pain and swelling of arthritis. The technicians are more engineers and computer experts than biologists because computer-controlled robots perform the tests. These are not movie robots with personalities but banks of machines that deliver carefully measured samples of the thousands of chemicals to a few million test cells in the 1,536 wells of rectangular plastic plates. Each well receives 5 millionths of a liter of a chemical solution, and each chemical is tested in triplicate at four different concentrations. The chemicals are stored in a computer-inventoried system, and the robots are programmed and instructed as to which collection to sample to make up the solutions to be delivered to the wells of the plates. The plates then travel along a conveyor belt to an incubator set at body temperature, where they are held for 4 hours. They then move, again as programmed, to a workstation, where a robot dispenses a series of chemicals to cause the contents to fluoresce with an intensity in direct relation to the amount of prostaglandin present. Now Lynne spends her time reviewing the computer-generated reports of "hits," chemicals that blocked prostaglandin production but did not simply kill the cells. Lynne also goes to meetings with managers who want to know which of the 20 or 30 hits she reported might be their next big drug. They worry about the complexity of each chemical because if it is too complex it might be expensive or even impossible to manufacture in large amounts. The managers also want to know about the strength of the patent position. Had anyone published a paper or filed a patent on a chemical that was similar? Lynne often asks herself whether this was why she went to graduate school—to baby-sit all those machines and make presentations of pretty graphs to a committee that grumbled if you found something and grumbled if you did not. Her thesis advisor, Ralph, had told her, when she told him at her thesis defense party that she was going to work here, that she would not like it; industry was not a good place for a biochemist with her ability. She would end up a drone, spending more time going to meetings than doing science. What committee, Ralph asked, discovers anything? But Lynne wanted to do something useful, to find medicines that help people, not just spend all her time writing grant applications. For what? To do the research and write papers, Ralph had said. But who was actually helped if all she had to show for her effort were a bunch of papers?

So here she was, two years later, with her own little research group, all those machines, and maybe a real drug, and now they grumble.

DRUG DISCOVERY SCENARIOS

People have disparate images of how new medicines are discovered. One image is that of the selfless scientist, working alone, who discovers the life-saving drug by chance. For example, many of us have heard the story of how Alexander Fleming returned to his untidy laboratory after a vacation and found a bacterial culture plate contaminated with a mold that stopped the bacterium from growing. Eureka! Penicillin! A more contemporary image for drug discovery is a completely different scene—Lynne's laboratory of large, dimly lit rooms filled with robots tended by a few technicians who push buttons and peer at computer screens. The stories behind the current contents of our medicine chests are less romantic than the Fleming myth and less automated than the image conjured by the companies marketing those robotic laboratories. But they are compelling stories of human effort and creativity.

Scientific and technological advances have transformed the methods used to discover new medicines. From the earliest times, healers have administered minerals or extracts of plants and animal parts to suffering individuals. These potions were both folk traditions, handed down from generation to generation, and the privileged knowledge of priests and shamans. With the development of the science of chemistry in the nineteenth century, the active principles of the potions could be extracted and purified, and the structures determined. These scientific advances launched the pharmaceutical industry, built in part on the first successful chemical industry, the dye industry. The first of the purified compounds was morphine, isolated from opium in 1806 (Porter 1997). Soon after, quinine was isolated from the bark of a tropical tree (Landau et al. 1999).

The development of the fields of physiology and pharmacology provided insights into how the various organs and systems of our body work and how drugs interact with those systems to bring about their effects, both desirable and unwanted. This led to the discovery of more and more active drugs, not only by identification of active principles from known natural sources but also by the synthesis of new compounds. Drugs to treat pain and fever were discovered. Vaccines and serums to protect against infectious diseases followed, based on a growing understanding of the basis of infectious diseases such as smallpox, diphtheria, typhoid, and anthrax.

The discovery of the first antibiotic, penicillin, by Fleming is often seen as a lesson of the value of accidents to the prepared mind. The popular story of a contaminated culture plate found after a vacation break is true. What is often not described are the half-century-old insights that preceded the discovery of the penicillin-producing mold and the organized international industrial effort it took to bring the drug to reality as a treatment for infections, an effort made urgent by the ongoing World War II.

In 1877, Louis Pasteur found that microorganisms themselves produce compounds that kill other types of microorganisms to rid themselves of competition for scarce nutrients. Pasteur's discovery was the reason Fleming was doing the work of looking for the antibiotic. Fleming's discovery in 1928 eventually yielded the first pure antibiotic, but only after an international team of scientists, driven by the need for treatment of infections from battlefield surgery, mounted a drive to develop a process to purify enough penicillin to treat a patient. The first patient was treated in 1942 (Goldsworthy and McFarlain 2002). Over subsequent decades, many medical antibiotics were discovered by testing large numbers of microorganisms from the soil and other sources for their ability to produce a molecule that killed other microorganisms. More effective and safer antibiotics were also found by applying both the growing understanding of how the body reacted to and broke down various antibiotics and the power of chemistry to modify the naturally occurring antibiotics to better target the structure or process that led to the microorganism's death.

DRUG DISCOVERY IN THE TWENTIETH CENTURY

During the first half of the twentieth century, chemistry, physiology, and pharmacology were joined by biochemistry, the study of the chemical reactions of living organisms, to provide a greater understanding of how the organs, tissues, and cells of the body worked. The components of cells and tissues that carried out the complex chemistry that supported life could be isolated and studied, and serve as suitable targets for a drug to stop this process or start that reaction. The general practice was to establish a test or assay for a particular process in the lab, using mixtures of the protein or cells that were thought to model the process in the body, and then test a series of chemicals for their ability to influence the test system.

Enzymes, the proteins of the body that do the chemical work, could be isolated, and their role in the body's functioning in health and disease pinpointed. Selected enzymes could then be used as test systems to discover new drugs by testing the ability of panels of chemicals to improve or stop the enzymes' functioning.

Other more complex studies of the workings of cells were also used. Researchers began to understand that the survival and functioning of cells depend on the control of the movement of calcium into and out of the cell, and they identified many different forms of calcium channels, membrane proteins that control the movement of calcium into and out of different types of cells. Each cell type could be used in the laboratory to find drugs that changed the way that particular cell controlled the entrance and exit of calcium. The identification of calcium channels was a scientific discovery that provided some of the earliest drugs to control high blood pressure. A drug that mimicked the effects of the removal of calcium ions from the medium bathing heart muscle cells suggested the potential of calcium channel blockers to dilate blood vessels and protect against organ damage in patients with high blood pressure (Landau et al. 1999).

Through the 1970s, such cell- and enzyme-based assays were the mainstays for drug discovery at pharmaceutical companies. But the science of pharmacology, the study of how drugs work, was becoming more sophisticated. As more useful drugs were found and more sought, the identification of new drugs resulted more and more from analyses of known drugs—their chemical structure and activities, side effects, and other properties. This follow-on approach draws on the expertise of chemists to identify the portions of molecules responsible for desirable or unwanted features of a drug and to redesign it to work better and with fewer side effects.

Compounds produced by the body were also identified, their structure solved, and their manufacture perfected to provide corticosteroids for inflammation, heparin to stop blood clot formation, hormones as contraceptives, and insulin for diabetics (Landau et al. 1999).

CANCER DRUGS

The discovery of drugs to treat cancer began similarly to the discovery of antibiotics, starting from an accidental observation and building through empiric screening, chemical modifications of successful drugs, and new insights into specific vulnerabilities of the target. The first use of a cytotoxic (cell-killing) drug for the treatment of cancers was reported in 1946 and was the result of the careful analysis of World War I military casualties. Nitrogen mustard had been used as a poison gas in World War I, and its toxic effects on the cells of the blood and bone marrow of gassed soldiers were reported in the medical literature.

Animal systems using tumors that could be transplanted from animal to animal and the ability to grow tumor cells in the laboratory not only provided tools for the testing of potential cancer drugs but also yielded insights that led to both the empiric discovery and rational design of such drugs (Papic 2001). The initial insight about cancer cells, modified and refined over the years, was that cancer cells are constantly undergoing cell division, free of the normal controls the body uses to regulate the process. Thus, the ability of nitrogen mustard to kill blood and bone marrow cells, cells that are normally more likely to be dividing than other nontumor cells, suggested that nitrogen mustard would kill tumor cells. Nitrogen mustard was tested and found active in animals with tumors and against human tumor cells in the laboratory. The question was whether a dose could be found that would kill a sufficient number of cancer cells in a patient without killing all the vulnerable normal cells, such as those in the blood and bone marrow and those lining the intestinal tract. This turned out to be the case when in 1942 the drug was tested in 67 patients with cancers of blood-forming organs, lymphomas, and leukemias (Goodman et al. 1946). Much of this work remained classified until after the end of World War II and was not reported until 1946 (Papic 2001).

Since that time, many more cancer drugs have been discovered by tests of large numbers of natural and synthetic compounds for the ability to kill tumor cells cultured in the laboratory. Also, successful drugs have provided leads for the design of new agents, based on the chemical structure of the active compound and more recently on the identification of the specific process or target in tumor cells that confers susceptibility to the agent.

As understanding of the biology, biochemistry, and genetics of tumor cells grew, it became clear that tumor cells acquired the ability to undergo cell division free of normal controls as a result of genetic changes, providing new drug targets. The biochemical processes that support or trigger cell division provide potential targets for cancer drugs. Families of growth regulator proteins signal normal cells to divide, and others signal cells not to divide. It has become apparent that the tumors have acquired sets of genetic changes that make the normal off switches for cell division nonfunctional. As a result, proteins and other molecules that trigger cell division drive the unrestrained growth of tumors. For example, epidermal growth factor (EGF) stimulates the growth of cells and tissues derived from the outermost layer of the embryo, including, skin, mucous membranes lining body cavities, and certain glands. Many tumors arising from these tissues rely on signaling from epidermal growth factor to survive and divide. They receive this signal through a docking protein for EGF on the cell membrane, called the EGF receptor. Erbitux, the cancer drug at the center of the ImClone legal battle, is an antibody, an immune system defense protein, that docks into the EGF receptor and prevents EGF itself from docking, so that cells, such as tumor cells, that rely on EGF signaling have their growth prevented and in fact die (Baselga 2002; Ciardiello and Tortora 2001). The antibody was purposefully developed using new molecular and culture techniques.

A similar science-based, rational process coupled with screening led to the discovery of the cancer drug Gleevec. A consistent finding in chronic myelogenous leukemia (CML), a form of leukemia with no known cure, is a genetic rearrangement between a region on chromosome 9 and another on chromosome 22, a rearrangement that leads to the fusion of the c-abl oncogene with the bcr gene. Oncogenes are normal genes that when mutated can cause a cell to become cancerous; that is, to lose the normal response to growth-control signals from the body. The protein that is synthesized as a consequence of the bcr-abl fusion was found, as an extension of studies in viral oncogenes that cause animal tumors, to have a particular biochemical activity, termed tyrosine kinase. The bcr-abl tyrosine kinase catalyzes the addition of phosphate groups onto the amino acid tyrosine in critical cell signaling proteins. This change causes the cells with the genetic rearrangement to become a cancer—to grow, spread, and survive free of normal regulation. Thus, searching for a biochemical inhibitor for the bcr-abl tyrosine kinase seemed a rational way to find a drug to treat CML. A group at Novartis set out to do just that. After finding some less than ideal leads by screening large chemical libraries for the ability to inhibit this particular tyrosine kinase, they made systematic modifications in the leads to find the most powerful specific inhibitor of the bcr-abl tyrosine kinase, Gleevec. Thus, this group built on insights into the genetic and biochemical changes central to CML and through screening and chemical lead optimization came up with a drug that warranted further testing (Druker 2002).

DRUG DISCOVERY AS AN INDUSTRIAL PROCESS

The biochemical targets that have been used for drug discovery are numerous, with more constantly being discovered as the body's complexity is further and further detailed. The draft of the human genome, the 3 billion base pairs and 25,000 or more genes (the number is in dispute), and the 100,000 different proteins our cells produce add to the challenge, as researchers try to assemble a picture of what each protein does and when, where, and how it interacts with other proteins. The complexity of the analysis has created an entirely new discipline, called bioinformatics, to try to tease understanding from the large bodies of biologic data being assembled by the analyses of genes and proteins.

Bioinformatics is also used for modern approaches to drug discovery. Large libraries (100,000 or more molecules) of chemicals, synthesized based on rational and step-by-step changes in structures or on random changes, are tested for the ability to react with sets of molecules chosen as promising targets for new drugs. Very small amounts of the target molecules are placed in many rows and columns on plastic plates or silicon chips and exposed to the chemical library to see which of the thousands of chemicals can interact with (i.e., stick to) the target. Various dyes are used to signal interaction. This process requires automation to distribute the samples for testing, robots to do the tests, and computers to assemble and analyze the results. Newer techniques allow many different compounds to be synthesized on small beads, which are then used in robotics-based assays to test for interactions with the target arrays on a chip. This industrialized version of discovery, called high-throughput screening, has been the direction pharmaceutical companies have taken to systematize the discovery process given the vast body of information that has been assembled about the chemical components that make up a human body. That this process will provide new and better drugs than the serendipitous discovery of Fleming or the biologically insightful discoveries that gave us drugs such as the contraceptive pill remains to be seen. A veteran pharmaceutical executive, Jurgen Drews, noting the small number of entirely new drugs since the mid-1990s, has suggested that the current high-throughput screening in the search for new drugs is hampered by a lack of focus on identifying and refining disease-specific targets for the screens (Willis 2004). For the most part, the individual scientist's eureka moment is an anachronism.

THE DRUGS PROVIDING OUR SCRIPT

We will track two drugs: Copaxone, a treatment for certain forms of multiple sclerosis, and Celebrex, Roger's arthritis drug. Each was discovered using different versions of more conventional methods of mid-twentieth-century drug discovery. The story of their discovery and development will illustrate both the power and limits of drug discovery using laboratory assays based on the prevailing understanding of the molecular and cellular basis of how the body works. These two drugs were selected because they can be followed from discovery to the patient—or, as the cliché goes, from bench to bedside—and illuminate the challenges inherent in the process whether the discoverer worked alone or as a member of a large team.

First, a note on names. During their history, drugs have several names: the specific chemical descriptive name, the company internal code name, a generic name, and a brand name. The first is based on the rules of chemistry, the last two issued by official agencies concerned with avoiding prescribing errors. The company's internal code name is just a shorthand way to refer to the drug before the generic name has been applied for and issued. Table 3.1 lists the various names for the two drugs we will be tracking. The chemical name will be given here for completeness but not used again! As we move from one name to another, from one stage to another, the name at the previous step will be provided along with the current name.

Copaxone—A Very Smart Guess

The discovery in the 1970s of cop-1, which would become Copaxone, a drug used to treat certain forms of multiple sclerosis (MS), resulted

Table 3.1
History of names for Copaxone and Celebrex

Name	MS drug	Arthritis drug
Chemical	L-glutamate copolymer with L-alanine, L-leucine, and L-tyrosine acetate	4-[5-(4-methylphenyl)-3-(trifluoromethyl)-1H-pyrazol-1-yl] benzenesulfonamide
Code name	Copolymer-1 or cop-1	SC-68535
Generic name	Glatiramer acetate	Celecoxib
Brand name	Copaxone	Celebrex

from the application of new chemical methods to basic research in immunology. It had become clear that the mammalian immune system makes very precise distinctions between proteins and that the ability to make this distinction in response to injections of proteins was inherited—that there was genetic control of the immune response. Michael Sela and colleagues at the Weitzmann Institute in Israel synthesized proteins for studies used to dissect how this genetically determined discrimination was brought about (Klein 1975; Silverstein 1989). This group then applied the power of the synthetic method to learn, using an animal model of MS, details of the process that led to the disease.

Multiple sclerosis is a disease of the central nervous system that is characterized by areas of loss of myelin, the fatty insulation of nerve fibers, in the brain and spinal cord. Nerve fibers missing myelin are not able to carry the signal from the brain to other parts of the body. This leads to the symptoms of MS, including difficulty walking, vision problems, weakness and fatigue, problems with bladder and bowel control, and painful muscle spasms. Circumstantial evidence supported the idea that the myelin damage in MS was caused by an inappropriate attack by the immune system on myelin, leading to inflammation and a failure of nerve function (NMSS 2003; Rolak 2003). Experimental models of MS in mice, rats, and guinea pigs had been developed by injecting the animals with brain-derived materials. Animals inoculated with extracts of brain tissue from other mice, rats, or guinea pigs developed a range of symptoms, including difficulty moving and paralysis of the legs and tail. The symptoms were similar, but not identical, to the symptoms of MS. For one thing, though MS in humans is not a lethal disease, some rats, mice, or guinea pigs treated with brain tissue extracts died. The animal model of MS is called experimental allergic encephalomyelitis (EAE), meaning inflammation of the brain as a result of the purposeful injection of material to cause an immune reaction. The evidence that the immune system was responsible for myelin damage in EAE was strong; immune system cells taken from an animal that had been treated with brain tissue would cause EAE in another animal if just immune system cells were injected (Sevach 1999).

The researchers at the Weitzmann Institute in Israel were trying to determine precisely what component of the brain extract caused the animal to mount an immune attack on its own myelin. Other researchers had separated the brain extracts into different proteins and attempted to cause EAE in mice with each. They were able to cause EAE with only one of the isolated proteins, myelin basic protein (MBP), but only if it was injected along with an emulsion of water and mineral oil, plus a detergent and heat-killed tuberculosis bacteria. These oil-water concoctions, called adjuvants, increase the ability of a substance to cause an immune response. If the myelin basic protein in only saline were injected, it would not cause EAE and would in fact reduce EAE symptoms if the animals were later injected with MBP in the adjuvant. Later, other protein components of myelin were also shown to cause EAE if injected with adjuvant. But at that point in time, myelin basic protein was the only protein identified with the ability to cause EAE (Teitelbaum et al. 1997).

The Weitzmann group wanted to define more precisely what the immune system was responding to in myelin basic protein. They did not know the sequence of the protein, but they did know which amino acids made up the protein—they knew the building blocks, but not their order in the protein chain. There are 20 known amino acids, some with acid side groups and some with basic side groups. Myelin basic protein is rich in basic amino acids. The Weitzmann researchers synthesized in random order three polymers of selected basic amino acids found in myelin basic protein, and tested these peptides to see whether they caused EAE or blunted MBP-induced EAE in guinea pigs. The basic synthetic polymers did not cause EAE in guinea pigs, but all were able to blunt the development of EAE if they were given after a dose of MBP known to lead to EAE (Teitelbaum et al. 1971). The same kind of experiment in rabbits with one of these polymers, called copolymer 1, or cop-1, gave the same results: EAE was seen in 19 percent of animals, down from 70 percent. These results formed the basis for a patent application in 1971; the patent was issued in 1974 (Teitelbaum et al. 1974a). Further studies of cop-1 in other animals, including monkeys and baboons, confirmed these results and will be described in Chapter 5. Cop-1 was on its way.

Celebrex—Snared in the Hunt

The identification of Celebrex was based on deepening insights of how a very old drug, aspirin, worked. Aspirin, acetylsalicylic acid, was a better-tolerated version of salicylic, a compound that had been used to treat pain and fever since the mid-nineteenth century, when the active principle of a fever- and pain-relieving bark was isolated (Porter 1997). The change to acetylsalicylic acid was made in an effort to make a better-tasting drug that did not cause vomiting (Landau et al. 1999). Aspirin was widely used for over a century, becoming the best-selling drug in the world despite the fact that no one knew how it worked. In 1971, Vane and others demonstrated that low concentrations of aspirin inhibited the production of substances called prostaglandins, which are manufactured by many cells and tissues in response to injury and are thought to be responsible for inflammation, fever, and pain. Aspirin was found to chemically modify and stop the activity of the key first enzyme in the synthesis of prostaglandins, an enzyme called cyclooxygenase (COX) (Insel 1996; Vane 1976, 1971). A whole series of other nonsteroidal anti-inflammatory drugs (NSAIDs) were developed over the following years in efforts to improve the efficacy of aspirin and reduce the side effects, particularly the damage to the lining of the stomach and intestines. A number were more effective, but all did damage to the gastrointestinal (GI) tract (Fries 1991). A series of discoveries in the 1990s indicated that there are actually two forms of COX: COX-1, present in all tissues, all of the time; and COX-2, produced in response to inflammation (Masferrer et al. 1992). The side effects of aspirin and other NSAIDs were thought to be due in part to the shutdown of COX-1 in the GI tract (Wolfe et al. 1999). A selective inhibitor of COX-2 enzyme was considered to have a good possibility of providing relief from pain and fever, with reduced or no damage to the stomach and intestines. A number of groups had identified selective inhibitors of COX-2, but none seem promising as improvements over the currently available NSAIDs.

Research groups from several companies, including Merck, Searle, and DuPont, began working very hard to find a drug that could be taken by mouth and was both a good COX-2 inhibitor and a poor COX-1 inhibitor. The available NSAIDs were a place to begin a systematic approach to adding and taking away chemical groups. The purpose was to learn three things: what chemical structures were necessary for COX-2 inhibition yet prevented COX-1 inhibition and allowed the drug to turn up in a sufficiently high concentration in the blood when taken by mouth. The latter characteristic was critical for the large potential market for a better NSAID, essentially a better aspirin or ibuprophen. The chemists at Merck synthesized a whole series of compounds based on one type of starting compound with some COX-2 selectivity (Prasit et al. 1999).

Celebrex (celecoxib) was discovered as the result of a purposeful search at Searle, a division of Monsanto, for a compound that would selectively inhibit COX-2. In 1989, Phillip Needleman, Ph.D., left his position as chairman of pharmacology at Washington University in St. Louis to become the head of research at Monsanto. He had been working on the biochemistry of inflammation, particularly the formation of cyclooxygenase, and strongly suspected that there was more than one form of COX and that a drug that inhibited one form but not the one in the stomach would be a safe and useful treatment for inflammation and pain. Finding a selective inhibitor became an important project in the Monsanto Discovery Group. Doctor Needleman had brought a team of researchers with him, and they recruited John Talley, Ph.D., and his Searle team of synthetic chemists to find a selective cyclooxygenase inhibitor (Talley 2003). The biologists had isolated the genes for the two enzymes and engineered them for production in cells cultured from insects (Gierse et al. 1995). The compounds the chemists synthesized could be tested to see whether they inhibited COX-2 but not COX-1. Starting with the chemical structure of some compounds that had anti-inflammatory activity but did not damage the stomach, Talley and his team synthesized and tested over 100 compounds, replacing different parts of the chemical structure with particular chemical groups to see which one gave the best preferential inhibition of COX-2. The compounds were tested in the lab on the isolated enzymes and in rats to see which ones had the best properties. Substitutions were made that were known from previous experience with many different drugs to allow the uptake of the drug from the stomach into the bloodstream and to allow the body to break down and excrete the drug within a few hours. The first characteristic was required if the drug was to be taken as a pill, and the second was necessary to make sure that the drug would not accumulate in the body and pose a danger.

From this series, the Searle group identified seven compounds that had the necessary selective activity on COX-2. These compounds would next be tested in mice to see how well the body would metabolize them. Those that passed that test would be further tested to see if they did indeed reduce inflammation and pain in test animals, a story for the next chapter.

REDUCTIONIST SCIENCE PROVIDES THE DISCOVERIES, BUT ONLY BIOLOGY CAN BE TRUSTED

The discovery of Celebrex and Copaxone reflects variations on the process common in the boom era of drug discovery, the period from the 1950s to the 1990s, when major advances were being made in the tools and methods available for what can be called a reductionist biology-based era. From the 1950s on, basic research in the biological sciences and in biomedical research was booming with postwar infusions of funding. This research was driven by a common conceptual thread, which held that understanding how the body functioned in health and malfunctioned in disease at a deeper and deeper level, from whole organism, to tissue, cells, and proteins, and finally to genes, would provide opportunities to intervene to maintain health and prevent disease. The entrance of many physicists into biologic research in the 1960s and 1970s contributed to this reductionist culture. Although the question was only occasionally bluntly asked whether the whole body was in fact a simple sum of its parts, that was the assumption. When predictions made from studies at the chemical or biochemical level did not always hold when applied to whole organisms, it was only a spur to dig deeper. Everyone understood that not all of the pieces were known and understood. The failures of drug candidates teach biologists in the pharmaceutical industry and in academic research, if they pay attention, that things are more complicated than they assumed, but the process does not provide a rigorous test of the reductionist assumption. Only drugs that are successful at the first step are ever tested at the next step; the failures go no further. The tendency of many companies to build on successful drug molecules to find their successors may be based on the acknowledged gaps in understanding.

The two stories of Copaxone and Celebrex also track the growing sophistication and biochemical refinement in the approach to drug discovery, on a honing of the reductionist approach over twenty-five years. The discovery of cop-1 in 1971 was built on applying new techniques to two ideas grounded in basic research but with a slim chemical foundation: (1) myelin basic protein would, if given under the appropriate conditions, blunt disease in an animal model; and (2) myelin basic protein was rich in basic amino acids. Making and testing the polymers in the animals was a good guess but risky nonetheless. At that time, the understanding of the basic processes underlying EAE and MS did not provide a molecular target to test in the lab. The only available, accepted test was the whole animal. The discovery of the set of selective COX-2 inhibitors in the mid-1990s was based on a systematic synthesis of compounds, adding this chemical group and taking away that. The test system was a pair of pure enzymes. There was no guessing. More recent research indicates that selectivity based on purified enzymes may be an imperfect predictor for the selectivity for COX-2 versus COX-1 in the body. Tests of the biochemical results of the enzymes using two separate assays with whole blood are more predictive and informative (Fitzgerald and

Patrino 2001). Yet for Searle and the other companies engaged in this work, moving ahead into animal models serves as a test of the hypothesis underlying this approach, that a potent, selective COX-2 inhibitor will relieve inflammation and pain with less risk of damage to the stomach.

4

Test Tube Results Are Not Enough: Animal Tests for a Drug's Utility

Compounds that hit in a test tube assay may trigger a smug smile or even a shout from a lab jock but are, particularly in today's factory labs, common as dirt. Just because the fluid in a culture dish changed color or a cancer cell died does not guarantee anything except more tests. That lab test was an isolated system with only one or two things going on, not an intact, living, breathing mammalian body with billions of cells and hundreds of thousands of different proteins working as needed. After a test-tube assay has produced a lead compound or set of lead compounds, the next step is the testing of the compounds in animals to see whether the drug is taken up and distributed throughout the body and whether the compound works in an animal model using an animal that either spontaneously or as a result of an intervention has a condition that mimics the human disease the drug is intended to treat.

The lead compound must behave as desired in the body, which means it must enter the circulation in order to reach the desired target tissue; an arthritis drug to reduce pain and swelling that does not get to the affected joints would be of little value. Although there are some general rules about the features of chemical compounds that make them able to be absorbed into the blood from the stomach or from an injection site, these rules must be tested because their predictive power is not great.

First, researchers must select how a medication should be administered. In general, patients and physicians prefer medicines that can be taken orally, but there are exceptions; for example, if a patient cannot swallow. Some potential drugs may not survive oral administration because the stomach is acidic or because the digestive system produces enzymes that break down proteins and other compounds. The chemists and biologists have to work together in the development of the leads to build into the molecule those features that will allow the drug to be stable and to be absorbed. Obviously, to track how the body absorbs the drug, they must also develop a way to measure the drug in the circulation in the test animals.

The second critical goal of animal testing is to find out whether the compound has the desired effect in an animal model of the human disease. If more than one possible lead was identified from the test-tube assays, animal testing allows the selection of the most promising compound. Whether there is one lead or several, animal studies provide the scientists developing the drug with a level of confidence that the identified lead has a reasonable chance of providing benefit to humans. The Food and Drug Administration, to balance the risk of testing a new drug in patients, generally requires such evidence before permitting human trials to begin. There will be a great deal of animal safety testing done later, before human testing may begin, but first the researchers and FDA scientists need to have some sense of the possible benefit before exposing humans to the unknown risks.

DIFFERING VIEWS OF THE USE OF ANIMALS IN RESEARCH

Animal research forms the basis for much of the progress in understanding and treating human (and animal) disease. The practice is ancient but became systematic and central to the study of the human body during the Renaissance. Philosophers from the sixteenth century on have debated the morality of animal experimentation, their arguments centering on whether animals felt pain and the moral status of animals as living, feeling beings. Although the discovery of anesthetics and their use in animal experiments might have been expected to quiet this issue somewhat, revulsion at the use and potential misuse of

animals for human betterment sustains a significant activism opposed to the use of animals in research. Biomedical researchers also
believe that the misuse and maltreatment of animals is horrific; they
also hold the prevention and mitigation of human suffering as high
goals. The research community's solution to bridge the gulf between
the protection of humans and the protection of animals is to ensure
that animals are used in as humane a manner as possible.

Animal cruelty laws have been enacted in many countries to prohibit
inflicting needless pain to animals. In the United States, laws regulate
animal experimentation to assure that the use of animals is necessary
for the particular research purpose and that the animals are treated
humanely. The Health Research Extension Act of 1985 provides the
requirements for the performance of animal experiments in all federal
agencies and at institutions that are recipients of federal funding
(NIH/OER). The Animal Welfare Act (AWA), adopted in 1966 and
amended several times, provides for the registration of all animal research facilities with the U.S. Department of Agriculture (USDA).
The AWA covers pharmaceutical and biotechnology companies that
do not receive federal research funds (Animal Welfare Act as amended
2002). Under both laws, animal research facilities must meet standards for the care of animals and are subject to inspections. These
facilities must also establish Institutional Animal Care and Use Committees (IACUCs) to review research protocols and procedures pro-
moting animal welfare.

As initially enacted and subsequently amended, the AWA covered
dogs, cats, hamsters, guinea pigs, rabbits, nonhuman primates, and
farm animals. The AWA does not cover mice and rats, which account
for 90 percent of animals used in research (Bishop and Nolen 2001).
Efforts by animal rights activists to include mice, rats, and birds in
the list of protected animals in AWA have met with strong opposition
from both academic and industrial biomedical research institutions
and organizations. In 1998, the group Alternative Research and Development Foundation filed a petition stating that the exclusion by
the USDA of birds, mice, and rats from AWA protection was capricious and arbitrary. The group subsequently sued the USDA to modify
the rules. In the out-of-court settlement, the USDA agreed to produce
in a reasonable period of time a set of rules to include mice, rats, and

birds. The process has dragged on, aided by the inclusion in the 2001 Farm Bill of a prohibition of the use of USDA funds for that purpose. The 2002 Farm Bill went further to stop the process, amending the definition of "animal" in the AWA to exclude birds, rats, and mice *bred for research* (APHIS 2002). Because mice and rats used in biomedical research in general, and pharmaceutical research specifically, are nearly always bred for research, the change in AWA has removed the earlier commitment for developing rules that might have required protocol review and USDA inspection of facilities using only rats, mice, or birds.

Despite this successful effort to preclude the application of AWA's requirements to mice and rats, cruel, inhumane treatment of experimental animals is seldom, if ever, in the interest of better science. Stressed, maltreated animals do not provide well-controlled, reproducible results. At most pharmaceutical companies, veterinarians whose professional standards would not allow such abusive activities manage the animal facilities. Public opinion, efforts by activists, and a growing sensitivity to animal welfare have increased efforts by research scientists and research institutions to find replacements for conscious, living animals in research, reduce the number of animals used, and decrease the extent of pain and distress experienced by research animals.

THE WHY AND WHAT OF ANIMAL TESTS

Although debates may rage in the media and classroom about the ethics and morality of using animals in research, the use of animal models of disease is rarely debated in a general sense in the biomedical literature. From a scientific perspective, only the specific link between the animal model and the particular human disease is subject to debate (Rubinsztein 2003). In other words, does this model of that specific human disease have validity? All the relevant features of tissue damage, symptoms, and disease course in the animal model may or may not match those of the human condition, but there must be a consensus among experts in the field that the animal model is sufficiently similar in important aspects. There is no guarantee that a treatment that is effective in the animal model will turn out to be effective in humans, but failure in an accepted model bodes ill. Compounds that fail in a widely accepted animal model will rarely move to human testing unless benefit is shown in another accepted animal test system and the rationale for confidence in the validity of the second model can be justified. Animal models are developed and may be widely adopted, then refined, and later possibly abandoned as understanding of the disease process deepens and new models are developed. Animal models of human diseases are also used in basic research as tools to better understand the human disease.

For applied and basic studies, researchers may use animal strains, most often mice or rats, that spontaneously develop some or all of the symptoms of human disease. Large repositories of highly inbred and mutant strains of mice and rats have been developed. Mice are the favored species for tests on certain diseases. For example, mutant strains of mice with defective immune systems are often used to test cancer drugs because they will not reject a transplanted human tumor. Because of their small size, mice are attractive for early testing of new drugs because small quantities of the drug are required. Rats, in part because they are larger and provide more tissue and blood for analysis, are favored for other studies, particularly in physiological studies related to circulation in the heart and lungs and studies of the nervous system. Animals may also be treated in a way that causes the specific characteristics of a human disease. A tissue or organ may be damaged by exposure to particular chemicals. Again, this may yield a model only as good as the current understanding of the human disease.

More recently, genetic tools and embryo manipulation have been used to render genes nonfunctional or to insert specific genes into the genome of experimental animals. This has been most widely used with mice and rats, though other species, including pigs, goats, and sheep, are also being genetically modified. The animals that have lost a functional gene are called "knockouts;" the animals with new genes are call "transgenic." The new gene may be a modified version of a gene from the same species or may be a gene from another species, including humans. Transgenic or knockout animals may be further bred with other genetically modified animals to achieve more genetic changes. In basic research, genetically modified animals may be used to test the role of a specific gene in the functioning of a specific group of cells or tissues. They may also provide models of human disease for the testing of potential new drugs based on current ideas about the role of particular genes and the proteins they produce in a disease.

In drug discovery, animal models are used to provide clues as to whether one drug or another might be more or less likely to work in the human setting. They are rarely perfect predictors, but increased confidence suffices. No one really can expend the time, effort, and resources to establish how well animal models predict success in human trials. Such an assurance of predictability of the animal test systems could only be gained if both drugs that worked and others that did not work in animal models were tested in humans. Testing compounds in humans that are not active in the accepted animal model is generally held to be unethical, placing humans at risk with little likelihood of success. Confidence in the animal models comes both from the similarities of the biology and biochemistry of the human disease and animal model and also from the history of success in human trials of drugs that were active in the animal models. An animal model may not be 100 percent predictive, but at least it provides some confidence that the risk of failure in human testing may be reduced. Confidence will be highest when the mechanism of the new compound is similar to that of a successful drug, particularly if the successful drug had also been vetted in the same animal model.

ANIMAL TESTING OF COP-1

In their early work on cop-1, the researchers at the Weitzmann Institute had no such track record of successful drugs to rely on. In 1971, there was no effective treatment for MS. Glucocorticoids, hormones isolated from the adrenal glands, were just entering medical practice in an attempt to speed recovery from the attacks that characterized the disease, but there was no effective treatment to prevent those attacks or slow the unpredictable though sometimes inexorable decline into a bedridden state that faced the newly diagnosed. Although lesions of myelin loss in the brain were seen when the disease was first described in 1860, the role of the immune system and the inflammatory process triggered by the immune attack on the myelin loss were still being clarified by basic research. It was such basic research that had led the scientists to the idea of making the synthetic polymers to mimic myelin basic protein in triggering the specific immune response to myelin.

The ability of one of those polymers to blunt symptoms in the guinea pig model was sufficient for a patent but would not suffice for beginning to test in human patients. Although they had similar results in mice and rabbits, animal data in a model that more closely resembled MS were needed. Those initial studies used animal models that developed complete and permanent paralysis within 2 to 3 weeks. In humans, MS is most often a chronic, relapsing, remitting disease; attacks occur one or more times a year; the symptoms of loss of vision, paralysis, and so forth resolve more or less in a matter of weeks, only to be followed a few weeks or months later with another attack of the same or different symptoms. A chronic animal model with attacks, resolution, and relapses was needed for the next test. The guinea pig was again chosen, but the test used young animals of a different strain and changed the way they caused the disease because it was known that these changes provided symptom severity and disease course closer to that of MS. The researchers injected two different doses of homogenized guinea pig spinal cord in a mixture of oil, water, and dead bacteria into the young guinea pigs. At the high dose level (less than one-tenth of a teaspoon), attacks occurred at 10–14 days, and five of the ten test animals died in the first attack. The five others recovered, but 60 days following recovery, those five animals relapsed, with two dying. Of the three remaining animals, two had a second relapse and died; the surviving animal had significant paralysis. A second group of twelve guinea pigs was pretreated with five injections of cop-1 before the injection with the spinal cord homogenate. Only five of the twelve guinea pigs in this group experienced an attack, and these attacks were mild and occurred 90 days after injection. After 111 days, all twelve animals were alive and only two showed mild to moderate symptoms.

Impressive though it is, this pretreatment study is not relevant to the human situation. Multiple sclerosis is rare and unpredictable; because no one knows whether he or she will develop the disease, pretreatment is not an option. Researchers therefore performed a third high-dose spinal cord homogenate study in which eight guinea pigs received cop-1 at the first sign of symptoms. All cop-1–treated animals showed improvement, the attacks were shorter, and all eight showed complete recovery. Six of the eight did relapse; one died. Three animals had no symptoms at 340 days. The researchers then repeated the treatment experiments using a lower dose of the homogenized spinal cord homogenate (less than one-twentieth of a teaspoon). One group received no cop-1, and the second group received cop-1 treatment at the first sign of symptoms. At the lower dose of homogenized spinal cord, with no cop-1, all animals had attacks at 10–14 days; three of ten animals died within 23 days. Three of the surviving seven experienced progressive deterioration and were killed for tissue studies; three had relapsing disease. When a set of ten animals that had received the low-dose homogenized spinal cord were treated at the onset of symptoms with 5 days of cop-1 injections, three relapsed at 1–2 months, but seven animals did not relapse and were symptom-free at 270 days.

THERE ARE BIG DIFFERENCES BETWEEN GUINEA PIGS AND HUMANS

This demonstration of the benefit of beginning cop-1 treatment when animals developed symptoms was very encouraging (Keith et al. 1979). The next step was a study in a species closer to man, rhesus monkeys. The study involved five monkeys; all were injected with myelin basic protein (from cows) in an oil-water, dead-bacteria mixture. Cop-1, in an oil-water mixture without the dead bacteria, was given to two monkeys when symptoms appeared, between days 16 and 24. The cop-1 was given daily for 15 days in decreasing doses. Three monkeys were controls; that is, they received no cop-1. Two of the three control monkeys received injections of the oil-water mixture without dead bacteria; one received no injection except the myelin basic protein. Both cop-1–treated monkeys improved after four or five injections and recovered completely by day 29. The three monkeys that did not receive cop-1 were severely affected and died by day 35. One cop-1–treated monkey experienced relapses and died by day 91. The other cop-1–treated monkey remained symptom-free and was sacrificed for brain studies. The brains of all animals were examined for the inflammatory signs typical of EAE and active MS. All of the control animals and the cop-1–treated animal that relapsed and died had typical brain lesions; the treated animal that recovered did not (Teitelbaum et al. 1974b).

The numbers were small, but the animal data supported consideration of full-scale testing in humans. Safety testing in animals, and development of a method to manufacture large amounts of cop-1, were two hurdles that needed to be overcome, but the proof of principle had been accomplished by 1979. Professor Sela and his colleagues still had only an indistinct sense of exactly how cop-1 worked (Keith et al. 1979). In fact, exactly how and why the immune system attacked myelin in MS was poorly understood and remains so today (NMSS 2003). The Weitzmann group also did a small test of cop-1 in three patients with advanced MS and saw no improvement, which they attributed to the advanced stage of the disease (Abramsky et al. 1977).

THE COX-2 INHIBITORS MOVE INTO ANIMAL TESTS

Twenty years after the initial work by the Weitzmann group, the sequence of events leading to the discovery of Celebrex reflected the fuller understanding of the biology and biochemistry of the target for the new drug. Searle, seeking a selective COX-2 inhibitor, had methodically gone about the testing of a large number of compounds in biochemical tests in test tubes in the lab. Next, to sort out the connection between the lead compounds' chemistry and their absorption and distribution, the researchers administered seven of the COX inhibitor compounds to mice either by injection into a vein in the tail or into the mouse's stomach through a small plastic tube. All the selected compounds were effective COX-2 inhibitors, but some were also COX-1 inhibitors. Two compounds remained in the circulation for 100 to 200 hours, too long to be considered safe. Five compounds had half-lives in the blood of 3.3 to 5.6 hours, the range of removal from the body considered acceptable (Penning et al. 1997).

Searle then tested eleven of the new compounds, chosen to represent various chemical features, plus three established NSAIDs for their ability to treat pain, arthritis, and inflammation in three rat models of symptoms known to involve COX-2 (Siebert et al. 1994). In order to cause inflammation and swelling, the researchers injected an irritant into the hind paw of rats treated orally one hour earlier with one of the 14 compounds. To gauge the effect on inflammation, they simply measured the size of the injected paws three hours later. The next test was for treatment of pain. Inflammation causes increased sensitivity to pain, which can be measured by testing how quickly the rats picked up the injected paw from a glass floor heated by a lamp. The third test was a rat model for arthritis caused by injection of mineral oil containing killed bacteria into one paw. Arthritis, as measured by swelling of the paw on the other side of the body, develops within 15 days after the injection. Treatment by mouth twice a day with one of the 14 compounds was started on day 15 and continued until day 25. For each study, animals received a range of doses of the compounds, and the compounds were rated by the concentration of drug that provided a 50 percent reduction in the symptoms of swelling, increased pain sensitivity, and arthritis. A lower dose providing 50 percent relief indicated a more powerful drug. Because the search was for a drug that, because of its COX-2 selectivity, would be safer for the lining of the digestive system, a study was done in rats to the determine the amount of damage to the stomach lining visible with a microscope. The COX-2–selective inhibitors caused little or no damage; all of the commercial NSAIDs and new nonselective COX inhibitors did damage the stomach lining. Based on the "overall biological profile," one COX-2 inhibitor, code number SC-58635 (generic name celecoxib), was selected (Penning et al. 1997).

THE SCIENTISTS HAVE DONE THEIR PART; NOW THE BUSINESS PEOPLE MUST MAKE CHOICES

Animal experiments have shown that cop-1 can reduce the symptoms of the MS-like disease and have allowed the selection of a COX-2 inhibitor, celecoxib, with the required beneficial effects without stomach damage. The next step requires a team to decide whether to move ahead with the development of the leads. Now the game is not just about science or even medicine but business—big business—so the decisions are not lightly made and are grounded in issues addressed in the next chapter.

5

The Business Decisions:
Committing to Development

Before they reach the consumer, all medicines must jump a critical hurdle—a company must make the decision to move a drug candidate from research to development. The studies needed to bring a drug to market are costly and time-consuming, and the management deliberations behind the decision to commit a company's resources of time, people, and money to the development of a new drug are not only critical to the success of the particular compound but also can be central to success of the company in general. The company must consider the costs of the further work needed to bring the drug to the market, the chance that the drug will be approved, and the chance that, if approved, it will sell well enough to recover the costs and make a profit.

Generally, at any one time, large pharmaceutical companies are considering several different drugs that have completed the initial proof of principle. These may be drugs for the same purpose, to treat the same disease or symptom, or they may be useful for different diseases or symptoms. Large companies invest in discovery research in those areas where they think a new drug is needed based on their experience plus advice from both company scientists and physicians as well as outside consultants.

Another way large pharmaceutical companies acquire drugs is by licensing them. These drugs may be compounds discovered by academic researchers, by other companies that may want to license the drug to others for development, or even by government laboratories that carry out substantial basic research but prefer to license potential drugs to companies for development. Licensing from a company, university, or government lab involves negotiations for the terms of the license, which may involve upfront payments, payments when certain milestones are reached, and royalties, which are a portion of the sales when the drug is approved for sale. The licensing staff submits each candidate they have found for scientific and business evaluation just as the discovery scientists do.

WHO DECIDES AND WHAT DO THEY CONSIDER?

When a compound has shown effectiveness in an appropriate animal model, and found to be taken up, distributed through the body, and eliminated in an appropriate amount of time, the compound is submitted for consideration for development. In a certain sense, the clinical scientists and business staff sorting through the compounds submitted for evaluation are the customers of the discovery scientists and the licensing staff. The clinical staff must see that they can put together and carry out the clinical trials needed for approval, and the business staff must see that they not only can sell the drug but also that it will be profitable.

The package of information that the research scientists submit to these decision makers includes not only the proof of effectiveness in the animal models of disease but also extensive intelligence on the condition for which the drug might be used, other drugs available for that purpose, and drugs being developed by other companies or studied by academic investigators for that condition. Companies invest significant amounts of time and money in tracking the success of other drugs on the market, and in researching the strengths and weaknesses of the competing pharmaceuticals. They also make every effort to keep abreast, through attendance at scientific meetings and analyses of industry publications, press releases, and other available information, of drugs being developed by other companies.

For this decision, company chemists and chemical engineers experienced in drug manufacturing must project how easy or difficult large-scale production of the compound would be. The scale of the production needed is estimated not only for the development work of animal safety studies, human clinical trials, and drug stability studies but also for the sale of the drug once it is approved. These projections are developed in collaboration with the preclinical and clinical scientists who estimate the amount of the drug needed for the development studies and the amount needed once the drug is approved.

The clinical scientists, usually physicians with training and experience in treating the disease targeted by the drug candidate, put together the clinical development strategy, which is the plan for the studies needed to show safety and efficacy. They estimate the dose, the schedule (how often a patient would take the drug), and the duration of dosing. With the business staff, the clinical scientists estimate the size of the projected market (how many people have the condition for which the drug is intended) and the projected market share (what fraction of those patients would likely take this drug). The business considerations include the unmet need or needs that might be met by this compound, the current and anticipated competition, and how much better the candidate would have to be to not only obtain approval but also gain a significant market share. These estimates are educated guesses based on an analysis of how much better this drug might be than the currently available drugs and competitors' drugs in development. The business analysis also addresses whether the drug would fit into the company's business; whether they have the ability to market the drug and have the sales staff to sell the drug.

The Number Crunching

The scientists and business analysts also project the cost of development. This includes not only the out-of-pocket expenses but also the cost of development. The money used to develop a new drug comes from the earnings of the company, earnings that stockholders expect to receive in the form of dividends. These retained earnings are, in a sense, being borrowed from the expected dividends of the stockholders, in anticipation of greater earnings in the future. The cost of retained earnings used for development includes the percentage return the stockholders would have expected if they had received dividends and reinvested those dividends in stock in the company over the years it took to develop the new drug. Capital costs also arise from the cost of issuing common stock, the dividends of preferred stock, and the rate of return of corporate bonds. The company's financial staff calculates where the company derives the capital needed for development—the fraction from retained earnings, from common stock, or from preferred stock and bonds—and multiplies each by the cost of the money, expressed as a percentage of the expected return from the company. This calculation provides a weighted cost of capital and is included in the estimate of capitalized cost—how much money it will take to develop the drug.

The numbers are large. Estimates have been made of the capitalized costs of developing a new drug (DiMasi 2001; DiMasi et al. 2003; Grabowski and Vernon 1990; OTA 1993). The estimates vary widely, from $125 million (U.S.) for the years 1970 to 1979, to $231 million for the years 1970 to 1982, to $820 million for drugs approved in the 1990s. All these studies are based on a variety of estimates and assumptions, and each has its own uncertainties, but there is little doubt that the cost is high and rising. Although we can question how accurate and forthcoming companies were in providing numbers to the authors of these studies, this is not an issue when a company sets out to calculate its own projected out-of-pocket expenses, capital costs, and estimated duration of development activities for a particular drug.

THE DECISION IS REALLY A CHOICE

Because they may have several candidates for development, the question facing the management of a pharmaceutical company is whether they should commit resources to one compound or another. They will compare uncertainties in early projections of costs for each candidate. These projected estimates of costs will be reviewed regularly during development to determine whether the risk for a particular compound has increased substantially enough to warrant dropping the project. The analysis is never over, but two key decision points are at this stage and later when the large, critical, and expensive clinical trials must be done.

The other half of the financial equation is the projected earnings from the drug. Remember this must often be estimated before a single human being has been treated. The price the company might charge for the drug is estimated based on currently approved drugs in the same setting, the potential advantages of the candidate compound, and its possible disadvantages, plus a realistic look at the issues that might surround insurance company reimbursement for the drug. We will address pricing and reimbursement in detail later, but at this point an experienced committee of company experts develops an estimated price for the drug and the best- and worst-case market size from the time of approval over succeeding years of its patent and exclusivity life.

Patents, Intellectual Property, Matter a Lot

Because the duration of the monopoly the company will have for the sale of the drug greatly influences the price, earnings, and therefore the return on investment, an important factor in the decision to develop a potential drug compound is the patent protection for the compound. It is widely held in modern industry, and especially so in the pharmaceutical industry, that the only way a company can recover the costs of discovering and developing new products and make a profit is to have a substantial period of marketing exclusivity. Drug marketing exclusivity, the period of time when only the sponsor of the drug, the company receiving approval from the FDA to sell the drug, is legally permitted to market the drug, derives from both patent law and certain FDA provisions of exclusivity. The FDA provides for various periods of exclusivity for drugs that treat rare disorders, drugs tested in and approved for children (CDER 1999), and drugs that are entirely new chemical entities. These extensions of protection from competition, included in several amendments to the law governing the FDA, are intended to reward and therefore provide incentives for innovation; for developing drugs for patients with rare disorders, where the market may be too small to provide the profit a company expects; and for performing controlled trials in children, if the drug is likely to be prescribed for children. But at the outset the patent itself, especially its claims, is critical.

Applications to the U.S. or foreign patent offices for a patent may include claims for the chemical compound itself, the form of the compound, use of the compound for a particular purpose, and the process by which the compound is manufactured. The compound itself, the substance of matter, is the most sought-after form of intellectual property protection for a potential drug because creative, motivated scientists may easily discover other methods of manufacturing and formulation, the exact makeup of the medicine a patient takes. After the patent application is submitted, the claims are examined by the patent office for novelty, nonobviousness, and utility. Previous public disclosure of the substance of a claim by the applicant or others, called prior art, will lead to rejection of the claim as not novel. The art and science of drafting claims that withstand examination are crucial skills for a company. Patent lawyers, often with advanced degrees in biology and chemistry, are valuable members of the discovery team. Prior to 1994 and the passage of Public Law 103-465 as the result of the United States' signing of the Trade-Related Aspects of Intellectual Property agreement, a U.S. patent was valid for 17 years from the time of issue. Currently, an allowed U.S. patent provides protection for 20 years from the date of filing (Uruguay Round Agreements Act 1994; U.S. Code 2003). The life of the patent is particularly significant for the pharmaceutical industry because it may take 10 to 15 years to gain approval for the sale of a drug, leaving only a few years of patent protection.

The impact of patent expiration on sales and revenue has been analyzed by the now defunct Congressional Office of Technology Assessment (OTA) (OTA 1993). They tracked the volume and revenue of sales of compounds relative to the loss of substance of matter patent protection in the years 1984 to 1987. The OTA report suggests that cash flow to the originator company approaches 0 by 20 years after patent expiration because of the development of new and presumably improved drugs for the same purpose and the entrance of generic forms of the drug, which can drop the value of the originator's compound to that of a commodity (OTA 1993). It is understandable that the duration of market exclusivity and the durability of that protection are critically important to the decision to develop a compound. If substance of matter protection is not possible, the ease of another company coming up with a way to manufacture, formulate, or use the drug reduces the attractiveness of committing to development.

The decision to begin development of a new drug is thus, ideally, a systematic process of evaluating the projected cost of development in time, resources, and people and the potential for the drug to obtain approval and sell sufficiently well to recover the capitalized costs of development plus a profit. The standard business method for assessing the value of an investment is to calculate its net present value (NPV), the current value of what the investment is projected to bring in minus the present value of the project's cash outflows, corrected for the cost of money. For many companies, the decision for any particular compound is not made in isolation. The candidate is ranked as an opportunity against the other candidates that are available from in-house research or from licensing. Generally, a company has limited resources and must assemble a portfolio of compounds to develop, a pipeline, that not only has the most promise but also leverages the expertise and experience available to the company. With all the uncertainties in the process of drug development, some large companies have begun to use a statistical approach to this decision called decision analysis, a method that allows a balanced weighing of the probability of success for each of the steps required to gain approval and to return profits to the company (Gittins 1996; Rodriguez 1998; Wang 2003).

These are decisions made by human beings and are associated with great risk, uncertainty, and potential reward for the company and the individuals. The process from the decision to the market is complex, technically challenging, and long, often ten years or more. There will naturally be many points when the risk of continuing may seem too high or when another opportunity presents itself. There may well be disagreements and differing points of view about the ease or difficulty of the scientific, production, or business tasks. Sometimes the compound, the medical need, or the technology finds a champion, someone with sufficient credibility within the company or outside who so strongly believes in this compound that he or she will tenaciously fight for the survival of the compound in development.

THE DECISIONS FOR COP-1 AND CELECOXIB—THE ROLE OF CHAMPIONS

The drugs we are tracking represent two real-world variants on the decision process, but each had a champion. Copaxone (copolymer-1), invented by Michael Sela and colleagues at the Weitzmann Institute, had undergone limited animal safety testing and small pilot human tests in Israel. A rigorously designed, double-blind, placebo-controlled trial led by Dr. Murray Bornstein, M.D., at Albert Einstein Medical Center demonstrated that 20 mg of copolymer-1 injected once a day for two years reduced the number of relapses in patients with the relapsing remitting form of MS. Michael Sela shared the as yet unpublished news with Eli Hurvitz, the CEO of Teva, an Israeli company that had become a world leader in the generic drug industry (Teva 2003; Pinchasi 2003). Hurvitz was interested in moving Teva into innovative drugs, and the promising copolymer-1, given Bornstein's results, seemed a good opportunity. The company did its due diligence, and Hurvitz and his team decided to sign a license for the drug, a license with a catch, which is not an unusual occurrence (Pinchasi 2003). The problem the Teva scientists identified was whether copolymer-1 could be reliably manufactured at the level of purity and scale needed to obtain FDA approval. Teva would attempt to manufacture copolymer-1, first in the lab and then in two factory runs. If they could produce material that met predetermined specs for composition and purity, they would proceed with development. Irit Pinchasi, Ph.D., had recently joined Teva after a postdoctoral research fellowship in neurobiology at the Weitzmann Institute. Her first assignment was as the Project Manager for copolymer-1. The experienced laboratory and process development scientists at Teva, working with the Weitzmann team, were successful in the production feasibility tests (Pinchasi 2003). It was a deal. In this case, copolymer-1 was presented by a respected scientific and social acquaintance of the company CEO as a promising opportunity to a company looking to move into new areas. And it came with solid evidence of efficacy and safety in a study that would soon be published in the *New England Journal of Medicine*, one of the most, if not the most, prestigious medical journals. The decision was clear once the practical hurdles had been overcome.

As described in Chapter 3, celecoxib, SC-58635, was discovered as the result of a management-mandated, purposeful search for a compound that would selectively inhibit COX-2. Doctor Peter Needleman had determined when he joined Searle as head of research that a selective COX-2 inhibitor would be a priority project at Searle. The only convincing involved was to convince John Talley and his team that the challenge presented by the biologists was an interesting one

(Talley 2003). Once a small number of lead compounds were identified in 1993, the decision was not *whether* to develop one but which one and how. A discovery and development team of pharmacology, clinical, business, process development, and intellectual property experts met and chose one compound as officially sanctioned for clinical development (Talley 2003). When the champion of an approach and strategy is the scientist in charge, the decision process is straightforward.

NOW THAT THE DECISION IS MADE TO DEVELOP THE DRUG, HOW ARE YOU GOING TO PRODUCE IT?

Company chemists and chemical engineers will have suggested methods for production of the drug compound as part of their contribution to the decision to commit to development of the compound, and they may have tinkered a bit to test out their ideas, but once the decision is made, the production team must develop a method to produce enough of the drug at an acceptable cost. That is the next step, addressed in the next chapter.

6

Production of the New Drug

The next step in drug development is the development of a manufac-turing process—a reliable way to produce large amounts of the drug and the drug product, the mix of the drug with other materials, called excipients, to facilitate manufacturing and make the drug stable and able to be given to a patient. The success of a potential drug hinges not only on success in animal safety studies and clinical trials but pro-foundly on the ability to manufacture the drug in sufficient amounts in a cost-effective way while meeting regulatory requirements. For early research studies, the compound will be synthesized in small quantities in small vessels at a conventional lab bench. This scale of production for initial testing, perhaps 1–2 grams, is inadequate for the amounts needed for large animal safety studies, human clinical trials, and fi-nally commercialization. The cost per gram of drug synthesized at this small scale also would likely make manufacturing unacceptably expensive.

DEVELOPMENT OF THE PRODUCTION PROCESS

The development of an industrial-scale pharmaceutical production process occurs in several steps based upon how much drug is needed and the stage of development. In the first phase of discovery research

and early development, about 1 kilogram (~2.2 pounds) is also prepared in the laboratory. During this time, chemists and chemical engineers of the process development group will begin to work on assuring
that large-scale production is possible. For animal safety studies and
very early clinical trials, about 10 kilograms is produced in a plant
scaled for this level of production. As development proceeds, the
process for large-scale production will be refined to assure that
commercial-scale production will be possible, cost-effective, and
ecologically sound.

The process development group aims "to transform the research
synthetic procedure into a plant process by performing the necessary
laboratory experiments to achieve the goal of an ideal process" (Repic
1998). The ideal is a synthetic process able to produce the final drug
ingredient in a safe, ecologically sound, reproducible, economical, and
high-quality manner that will meet regulatory requirements. To be
safe, the reaction must be controllable within the reaction vessels. The
toxicity, flammability, and explosiveness of the chemicals required for
the synthesis are also considered. For ecological soundness, solvents
and chemicals are used sparingly and recycled if possible, and toxic
solvents are avoided. The process must suit the selected plant in terms
of scale of production, the ease of movement of reactants from vessel
to vessel, and the timing and nature of the chemical processes required for the synthesis (Repic 1998).

Controlling expense, a major issue in industrial-scale production,
is achieved by reducing the number of steps from starting material to
finished powder ingredient to be formulated into the drug. This is accomplished by improving the yield from each reaction step and by
maximizing the size of each batch to save labor and overhead. One
author has developed the concept of "atom economy, the maximizing
of the number of atoms of all raw materials that end up in the final
product" as a goal of good process development (Trost 1991).

Pharmaceutical chemists working on the development of the production process are keenly aware of another potential cost—the cost
of using a chemical step in the synthesis that is covered by a patent
held by another company (Cabri and Di Fabio 2000). The goal is a
process free of such requirements; all the better if the manufacturer
can patent protect the process.

The design and development of a manufacturing process for the chemical compound that is the candidate drug begins with a systematic analysis of the chemical structure of the compound. The chemists who first synthesized the drug in the laboratory, who were generally trained as synthetic organic chemists, were not necessarily worried about the suitability of their synthetic method for large-scale manufacturing. They needed to produce a sufficient amount of sufficiently pure material for laboratory tests and the initial studies in small animals. The process development chemists have a different task. They must mentally break down the synthesis of the compound into a series of chemical reactions they know can be achieved in a factory. The production of drugs is a type of fine chemical manufacturing. Fine chemical manufacturing equipment carries out sets of established chemical processes, adding certain structures, putting in an oxygen atom here or a hydrogen atom there and so forth. For factory production, the process development chemists select from the library of possible processes and put them in sequence. They begin with purchased starting chemicals, for example the amino acids for cop-1 or simple organic compounds for celecoxib, and finish with the chemical compound that is the active substance of the new medicine. The process development chemists may or may not use the same steps as the discovery synthetic chemists. They may have to develop an entirely new scheme of steps based on what can be accomplished safely and economically in the factory. They calculate the heat and pressure generated by the reactions and determine whether the factory equipment can withstand the stress. They determine the volume and stability of the solvents needed for each step to make sure explosions and fire are not a risk. They develop theoretic models of all the reactions and test their models in small versions of factory equipment, measuring heat, pressure, and yield at each step. If the synthesis goes on in different steps, they determine that the product of one reaction can be transferred to the next reaction vessel in the factory. In laboratory syntheses, the chemist may manually transfer the intermediate from one reaction flask to another. This will not work in a factory setting, so the process and equipment must be set up so that the production goes on without interruption. The process development group addresses all of these issues in terms of practicality, safety, cost, and patents. They rely on mathematics, chemistry, engineering, project planning, safety engineering, and environmental assessments (Tirronen and Salmi 2003).

REGULATION OF PRODUCTION

The manufacture of pharmaceuticals is, not surprisingly, highly regulated. In the United States, the Food and Drug Administration (FDA) has developed standards called current good manufacturing practice (CGMP) to be sure that all drug products (the pills we actually take) and the components (both active and inactive ingredients) have the identity, strength, purity, and quality that the manufacturer claims. The CGMP regulations provide standards for ten areas: organization and personnel, building and facilities, equipment, control of components, drug product containers and closures, production and process controls, holding and distribution, laboratory controls, records and reports, and returned and salvaged drug products. The regulations are grounded in the goal of assuring "that the drug products produced have the identity, strength, quality, and purity they purport or are represented to possess" (CFR 1994). The manufacturer is responsible for assuring that the drug product is not mislabeled or adulterated. Thus, the regulations require that the manufacturer provide people, facilities, controls, and procedures that are adequate to achieve a product that is neither mislabeled nor adulterated.

The quality control unit, a critical part of CGMP standards, is a team in a manufacturing facility that is charged with making sure that the manufacturing is done in compliance with the CGMP. The staff that performs and supervises the manufacture of the drug and the preparation of the drug product must have adequate training for their assigned tasks. Procedures and standards must be written down, including procedures for the receipt, storage, handling, testing, and approval of components used in drug manufacture. Similar procedures must be established for containers and closures—the bottles or vials and the bottle tops. Equipment must be designed, used, and maintained according to written procedures that ensure the quality of the drug product. Control methods, the ways that the drug, the containers, closures, and labels are sampled during the manufacturing process and are then tested and approved or rejected, must be written down and reviewed by the quality control unit. Equipment used for testing must be calibrated at suitable intervals and tested to make certain it is functioning correctly.

Documentation of all these CGMP procedures, the standard operating procedures (SOPs), is extensive and expensive, reportedly amounting to an average of 1,250 CGMP-required SOPs per company and 10–15 percent of total operating costs (Kieffer 2003).

WHAT THE PATIENT SEES AND RECEIVES
IS MORE THAN JUST THE DRUG

The drug chemical is often just a small part of the pill or injection a patient receives. Many drug chemicals are so potent that only a very small amount is needed to have an effect. The fine print on the back of even the aspirin or vitamin bottle lists inactive ingredients, which does not mean useless. This is not a rip-off; the other chemicals, called excipients, are necessary. Some provide bulk so that the pill is large enough to be manufactured, bottled, shipped, and seen. Other chemicals are included to help a pill disintegrate by swelling once exposed to a liquid—the water taken with the pill and the fluid inside the stomach. If the drug is to be released over an extended period of time, the pill may include chemicals that bind the drug and slowly let it go into the bloodstream (Jackson et al. 2000). If the pill is a tablet, binders and lubricants may be included so that the tablet can be formed in a machine and released from the pill-forming molds. The same combination does not work for every drug, and the process of coming up with the right combination of excipients for a particular drug chemical is called formulation. Injected drugs must also be formulated whether they are manufactured and shipped in the liquid form or shipped as a dry powder to be dissolved just before injection. In addition to water, the liquid formulation may include salts to control the pH of the solution, chemicals to prevent the drug from breaking down, and other chemicals to prevent the drug from sticking to the inside of the bottle or the syringe.

The drug used in animal safety studies described in Chapter 7 must be manufactured and formulated identically to the form used in the human trials. Changes in manufacturing or formulation may occur as the human trials proceed, but such changes will often require further, though limited, animal and human studies to show that the manufacturing changes do not alter how the body takes up and breaks down the drug.

Copaxone for injection is formulated with a chemical to improve stability and was originally supplied as a powder to be dissolved in sterile water approved for injection. Subsequently, it has been supplied in prefilled syringes. Celebrex is supplied in gelatin capsules with added chemicals to help the capsule rupture, improve drug stability, and increase its ability to dissolve and enter the bloodstream (Pfizer/Pharmacia 2002).

WILL THE DRUG BE STABLE IN THE WAREHOUSE, THE PHARMACY SHELVES, AND THE PATIENT'S MEDICINE CABINET?

Procedures for stability testing of the drug, determining how long the drug can be stored under a defined set of conditions without breaking down chemically, must be developed and written so that proper storage conditions can be provided for the drug at every step of the way from the factory to a patient's medicine cabinet. The results of all these tests must be recorded and reviewed. The bottom line is that there must be a written procedure for each step and each piece of equipment, and evidence that the procedures were followed must be documented and signed off by supervisors. If it is not written down, it did not happen.

HOW CAN WE BE SURE THE RULES ARE BEING FOLLOWED?

Quality control activities, systems, personnel, and practices to assure and document the consistent purity and quality of the drug compound and the formulated drug product require significant resources. A survey of six large pharmaceutical companies found a ratio of quality control personnel to manufacturing personnel between 1 to 4 and 1 to 7 (Anonymous 2004). Quality auditors examine lot records for proper dates and signatures and examine outgoing materials for proper labeling.

Before a drug may be sold, the FDA staff inspects the manufacturing facilities. The CGMP standards are further enforced by unannounced inspections of drug manufacturing facilities by FDA district office staff. The FDA inspectors check each aspect of the manufacturing process described in the CGMP. A drug that is not manufactured in conformity with the CGMP may be deemed adulterated by the FDA and the manufacturer subjected to fines of up to $100,000.

COULD IT BE DONE BETTER?

In 2002, the FDA announced an initiative to apply a risk-based approach to regulating drug quality, to encourage early adoption of new technological advances, facilitate application of modern quality management techniques, including implementation of quality systems approaches to all aspects of production and quality assurance, and encourage implementation of risk-based approaches to focus both industry and FDA attention on critical areas (CDER 2003b). The risk-based approach is intended to improve the efficiency of both the inspected (the manufacturer) and the inspectors (FDA staff) by using modern techniques to rank the processes and steps by the risk that failure in each poses to the consumer—the patient—and apply proportional scrutiny (Clinton 2003). To speed drug approvals and reduce cost without increasing risk to the consumer, the FDA is taking a page from the semiconductor industry to encourage quality improvement in pharmaceutical manufacturing (McCellan 2003). What this will mean for the pharmaceutical industry is not entirely clear to either the industry or the FDA, but both are pressured to do more with fewer resources without doing harm, and the FDA is working hard to get everyone concerned to buy into and focus on quality and quality improvement (CDER 2003b; Clinton 2003).

The production process may change as the development research proceeds. However, the FDA does require that the process-control procedures and documents of the drug used for animal safety tests and initial human testing provide evidence of the identity and purity of the compound. If the nature and quality of the drug used in animal safety studies are different from those of the drug used in human studies, how can the animal data be used to assure that it is reasonably safe to begin human dosing?

The two drugs we are tracking posed different manufacturing challenges—but I cannot tell you about them. The specific manufacturing processes are closely held secrets of each company. Details of all the laboratory tests and animal safety studies for any approved drug are freely available from the FDA, either directly on their Web site or through a Freedom of Information Act request, but details of manufacturing processes, though reviewed by the FDA and the subject of inspections, are redacted from any FDA documents. The small-scale synthetic methods and the principles of the syntheses are available in the patents and published papers for copolymer-1 and celecoxib, respectively, but what goes on in the factories is an industrial secret (Penning et al. 1997; Teitelbaum et al. 1974a).

NOW IT IS TIME TO SEE WHETHER IT IS SAFE TO BEGIN HUMAN TESTS

Once the production process is established, at least on a scale large enough to do animal safety studies and to begin to expose humans to the drug, the company begins the animal safety studies needed to convince themselves, the FDA, and others that it is safe to start human trials.

7

Laboratory and Animal Safety Testing

The next step for the drug on the path to the market is the performance of laboratory and animal safety and pharmacology studies. The laboratory and animal safety testing provide an understanding of the toxic effects of the drugs; the pharmacology studies describe both the effects and mechanisms of action of the drug and how the drug is taken up, distributed throughout the body, broken down, and excreted. These studies provide assurance to the company, the FDA, and the investigators who will do the human studies that the introduction of the drug into humans in the first clinical trials can be done with a minimum of risk. Because animal studies may take a long time, planning and performance of the studies required to begin human testing starts as soon as a decision is made for development, while the production process is being developed and tested. Many other laboratory and animal studies will be performed as human trials proceed, but the challenge now is the specific preclinical studies required before a single human is tested.

HOW WILL THE WORK BE DONE?

All of the work submitted to support the application for approval to market a drug, including these animal safety studies, must follow

FDA guidelines that describe both the kind of studies that must be done and how they will be done and documented. The rules about *how* critical animal and laboratory studies will be done and documented are outlined in regulations called current good laboratory practices (CGLPs) (FDA 1987), which were first developed in the late 1970s after FDA inspectors visited laboratories, including Searle's, where preclinical toxicology studies had been performed. The inspectors found that the studies had not been done in conformity with acceptable practice and that the problems were "so severe that in many instances the studies could not be relied on for regulatory decision making" (FDA 1984). Senator Edward Kennedy and the FDA raised public concern about this issue (Robinson 2003). Within a few months, Searle had developed a guidance document for these evaluation research activities and submitted it to the FDA and the Pharmaceutical Research and Manufacturers Association (Robinson 2003). In August 1976, the FDA released a draft GLP, based on the Searle document, and in 1978 it published the final CGLP regulations that describe the requirements of the facility, personnel, the handling of the materials to be tested, and the quality control and maintenance of the equipment used in the study (FDA 1976, 1987). The regulations, modified in the ensuing years, require written standard operating procedures (SOPs) for routine tasks and a written protocol for each study. CGLPs also include rules on record keeping and the format for reporting results. Every aspect of the study must be documented, down to records for the temperature of the refrigerator where samples are stored. This is a level of attention to detail and record keeping far beyond the normal functioning in an academic or industry research laboratory. The FDA inspects all CGLP laboratories for compliance with all aspects of the CGLPs, and if evidence is found that an FDA-licensed CGLP laboratory is not complying with any element of the CGLPs, it can lead to disqualification of the lab and delay or prevent approval of compounds undergoing testing (FDA 1998a).

THERE WILL BE CHECKS

Full compliance with CGLPs requires that an independent quality control team audit the study and document that the predetermined specifications of instruments, materials, and study results are met. The quality assurance group, though often employees of the company, must be administratively independent of those responsible for performing the preclinical and clinical studies. The CGLP audit team is just one of the quality assurance systems that the pharmaceutical company must have in place to support an FDA application for marketing. Quality assurance audits and reports are a constant element of the work that will be done to support approval of the drug (Robinson 2003). Questions or concerns at the FDA about the reliability of the data can slow or even stop the process. The study director, the qualified, responsible scientist, plans out and then signs off on each study before it starts. The facilities, equipment, personnel, and process must meet CGLP guidelines. When the study is completed, the study director must sign the report, and the whole process must be audited and reviewed by the quality assurance team. This is not a simple process, but leaving nothing but the actual results to chance provides the needed level of confidence in the results.

WHAT NONHUMAN SAFETY AND PHARMACOLOGY STUDIES MUST BE PERFORMED?

The CGLPs are the rules on *how* these nonclinical studies will be done. What are the rules on *what* studies will be done? The FDA had until the early 1990s a general set of guidelines for the testing requirements for nonclinical studies. They were in the process of revising them when it became apparent that a similar exercise had begun in the European Union in an effort to harmonize regulations of the EU member countries. Harmonization of European rules on the studies required for approval of a drug was driven by the desire to avoid the expense and delay of having to do slightly different nonclinical and clinical studies for each member country in order to gain approval for sale of the drug in each country. If a company only met the regulatory requirements of one country when studies were done, approval in another country would mean more studies, more costs, and more delays. If the rules were more or less the same in all of the countries making up the largest portion of the European market for prescription drugs, then one study would suffice for all countries. With that logic, extending the harmonization to the major non-European markets for drugs, the United States and Japan, seemed a smart move to all concerned. Given the growing cost of developing drugs, pharmaceutical companies in the United States, Europe, and Japan were beginning to think about efficient drug development on a global scale to reduce duplication, reduce costs, and raise profits. In 1990, a meeting was held among the regulatory agencies from Europe, the United States, and Japan at which it was agreed to develop a plan to harmonize the guidelines for safety, quality and efficacy of new chemical drugs (ICH 1997a). The International Conference on Harmonization (ICH) has developed guidelines for many aspects of the testing of new drugs. When approved by the ICH steering committee, after a process of drafting, consultation, and review, each set of guidelines is published as a guidance document by the member countries.

The current guidelines for nonclinical safety studies for the conduct of human clinical trials for pharmaceuticals, developed by the ICH and published in 1997 by the FDA, describe the required nonclinical (laboratory and nonhuman) safety studies and when they should be completed in relation to the clinical development (CDER 1997). These require that animal and laboratory studies go beyond the studies that provided scientific evidence that the compound might be useful for a particular disease or condition and do two things: (1) provide assurance that it is reasonably safe to begin testing in humans; and (2) support the selection of a dose to begin human trials. The results of the studies are provided to the FDA in a document called the investigational new drug (IND) application, which is required to begin human trials.

Pharmacology studies in animals provide information about how the compound affects the body, how the drug works, and the type and duration of responses to different doses of the drug. A standard battery of laboratory tests is also performed to assess the potential for the drug to cause genetic damage to cells, which might lead to cancer. This assessment of "genotoxicity" includes laboratory tests with bacterial and mammalian cells to see whether the drug causes genetic changes or damage to chromosomes. The blood and bone marrow of mice treated with the drug are examined for specific changes in the nucleus of certain cells, an accepted test system for genotoxicity.

Toxicology studies provide information about damage to any organs of the body, the relationship of damage seen to both the dose and how long the drug remains in the body, and whether the damage is reversible. The toxicity studies will start with "acute" studies, which are tests of a single treatment with increasing amounts of the drug. Then, depending on how many doses clinicians expect will be given to humans, animals are given repeated doses; these are called subacute, or sometimes subchronic, dosing studies. Both acute and subacute studies need to be performed in two different mammalian species; one may be a rodent (mouse or rat), and the other must be a nonrodent, usually dogs. Hundreds of animals may be used in all of the safety toxicity and pharmacology studies, but for each dose and schedule the group may be only three to five animals of each sex. The aim is to use as few animals as possible to develop scientifically credible information. Researchers design each study, including the size of the test groups, based on experience with similar compounds.

The data from the acute and subacute preclinical toxicity studies are used to establish a safe starting dose for human trials and to develop strategies for monitoring the subjects in clinical trials. For example, if the compound has effects on the kidneys of the test animals, the human trial protocols will direct investigators to pay particular attention to the status of the kidneys in the human subjects, possibly including more frequent blood and urine tests to check on how the kidneys are working. Other organs and tissues are not ignored; there is just added attention to suspected prime targets of the drug.

Although the animal safety studies, done at wide ranges of doses of the test drug, help determine a safe dose to use to begin human trials, the calculations used to define the safe starting dose are complex. Because test animals are, in general, smaller than humans, the doses given to the test animals are proportionally less than the amount clinicians expect to give to human patients. One important consideration is how the relative sizes of the test animals and humans are measured. It might seem logical to use weight to scale the animal and human doses, but the entire body's exposure to a drug, whether given by injection or by mouth, is a result of the entrance of the drug into the bloodstream and its movement through the body via the circulation. The surface area of the body, because it more accurately reflects the volume of blood, provides a better sense of the entire body's exposure to a drug as the drug is taken into the circulation and distributed to organs and tissues.

Prior to the development and 1997 adoption by the FDA of the ICH guideline on nonclinical safety studies, acute toxicity studies required increasing the amount of drug given to test animals until a dose was reached that resulted in the death of the animals. Researchers used the results to define an LD_{50} and an LD_{10}, which are the doses that result in the death of 50 percent and 10 percent, respectively, of the animals. To provide a safety factor, the human starting dose would be a small fraction of the LD_{10}. Now the ICH guidelines for nonclinical safety studies recommend a stepwise process to support the stages of human trials. As before, prior to beginning clinical trials, acute safety studies and subacute studies designed to define a safe starting dose in humans must be completed, but current guidelines suggest the highest starting dose in humans should be based on the no observable adverse effect level (NOAEL), the highest dose level that does not produce any adverse (toxic) effect in the test animal species that is most relevant to humans or most sensitive to the toxic effects of the drug. To help make the process more consistent, the FDA in late 2002 published a draft guidance document that provides a detailed process describing how to use the NOAEL to define a maximum recommended starting dose. The first step is to define the NOAEL from all of the animal studies. The NOAEL is then converted to a human equivalent dose (HED), generally by surface area, and the HED is corrected by a safety factor to identify a safe starting dose in humans. In July 2005 the FDA released guidance on estimating a safe starting dose. Information on a chemically similar compound in animals and humans, considering both adverse effects and blood levels of the drugs, may also be used in developing a safe starting dose.

The IND must include both the results of animal safety and analyses of blood and tissue levels of a drug in relation to adverse effects to support the proposed initial human studies. For the IND, integrated summaries of the toxicologic effects of the drug on animals in acute and subchronic studies and in test-tube studies are sufficient. Full final quality assured reports of each study, required for final submission of the application for approval of sale, might not be needed for the IND.

Prior to approval for sale, the drug will also be tested in long-term studies in rodents for the ability to cause tumors, called carcinogenicity. The rules for these studies have been the subject of some contention. The FDA's guidelines until the ICH addressed the topic indicated that a new drug intended to be used in humans for more than three months, with some exceptions for anticancer drugs, must be tested for carcinogenicity in both rats and mice for two years, just about the normal lifespan of the animals (two years for mice, three years for rats). The carcinogenicity studies begin with several large groups of animals who are treated daily with either a control substance or one of several different doses of the test drug, including the highest dose that the animals can tolerate; at regular intervals, animals in each group are killed and their organs and tissues examined for evidence of tumors. Carcinogenicity studies may take three years to run and analyze, cost a million dollars, and use a large amount of drug. The ICH review of the carcinogenicity testing requirements of the member agencies in the United States, Europe, and Japan led to an analysis of databases of carcinogenicity tests in rats and mice for a wide range of compounds, not just pharmaceuticals (Contrera et al. 1997). The question they asked was whether long-term carcinogenicity testing in both rather than one rodent species would be more likely to predict the potential for causing cancer in humans. The analysis did not support the need for long-term tests in two rodent species. The ICH guideline as published by all three agencies was a compromise, suggesting long-term tests in one rodent species plus, based on the scientific information about the drug and its interaction with cells, a 3–6-month study in an animal strain genetically modified to be more likely to develop tumors (ICH 1995). At least one commentator has written that this compromise was reached, despite the FDA's resistance, in order to speed the approval of new drugs at a reduced cost and posed a risk of harm to patients (Abraham and Reed 2003). After FDA published the ICH guidance in 1997, FDA staff scientists published articles, in their own names and not as spokesmen for the FDA, supporting the science-based flexibility of the ICH guidance yet describing the process of case-by-case negotiations between the FDA and the drug company to determine an acceptable set of studies for each drug (Contrera et al. 1997; DeGeorge 1998). As we will see, both celecoxib and copolymer-1 were subjected to two-year carcinogenicity studies in both rats and mice.

To pick up possible effects on reproduction and fetal and postnatal development, drugs are tested in rats, and possibly rabbits, by dosing animals prior to and during pregnancy and then following the effects through the next generation's reproductive cycle. These studies will generally be done during the period of human testing and, depending on the nature of the drug, human trials may not include testing of women of childbearing potential until the tests are complete. In some cases, trials in women of childbearing potential may be included prior to animal reproductive testing if the protocol requires pregnancy testing and the use of a highly effective form of birth control during the term of the trial. Tests in pregnant women and children are not begun until all animal reproductive studies are complete and safety data from human studies are available (ICH 2000).

For approval, the sponsor must also report studies of how test animals and humans take up, distribute, and eliminate the test compound, called pharmacokinetic and ADME studies (for absorption, distribution, metabolism, and elimination). If possible, animal ADME studies are done as the early human trials are being done so that toxic effects in animals and humans can be related to the timing of the appearance of the drug in the system after dosing and to similarities or differences in the way humans and test animals absorb, distribute, metabolize, and excrete the drug.

The ICH and FDA guidelines are just that—descriptions of the current best thinking of the FDA and ICH scientists. The studies performed for any particular drug may deviate from the guidelines for a number of valid scientific reasons. The studies that must be completed before human trials may begin may require more than three years to complete and document.

Not all currently marketed drugs have undergone the currently recommended animal safety testing. Drugs that were approved over ten years ago would have had different standards applied. There also may be timing issues, particularly for drugs that had a long history of development, including testing in human subjects. This is the situation with copolymer-1.

ANIMAL SAFETY STUDIES OF COPOLYMER-1

The early nonclinical safety studies performed with copolymer-1 were done to support the very early pilot human trials in Israel and also a small university-sponsored trial in early 1982 in the United States (Bornstein et al. 1987; CDER 1996). Acute, single-treatment studies were done in rats, mice, and dogs. These included injecting a very high dose (1,500 mg/kg or 3,300 mg/lb) into the veins of rats and observing the animals for 14 days. To put the animal studies in context, it is worthwhile to know that copolymer-1 was approved at a dose of 20 mg and that the weight of an average adult human male is about 154 lbs, or 70 kg (1 kg = 2.2 lbs). Therefore, 20 mg in a 70-kg man is roughly 0.28 mg/kg. No deaths or evidence of toxic effects were seen at this dose, which is nearly 4,000 times the approved human dose by weight. Mice were injected once into muscle with doses ranging from 100 to 2,500 mg/kg and observed for 14 days. Again, no effects were seen at any dose. Two dogs received 100 mg/kg, half under the skin and half into muscle, and were observed for 48 days, followed by extensive postmortem examination of organs and tissues.

Subchronic studies of copolymer-1 were done in rats and dogs. A group of ten rats were treated with copolymer-1 by daily muscle injection of 250 mg/kg for 3 months, followed by 3 months of daily injections under the skin of 200 mg/kg each day. No deaths were seen, though some swelling at the sites of injection and small changes in the bone marrow were observed. Ten dogs were given 10 mg/kg under the skin for 36 or 90 days. This study was incompletely reported in the final application, except to comment that no toxic or tissue effects were seen. These non-CGLP studies were all done in the early to mid-1970s.

Now a company seeking to develop a drug, also called the sponsor, will file an IND with the full complement of required preclinical studies, performed under CGLP. Generally, representatives of the company will meet with scientists and regulators from the FDA to develop the menu of animal safety studies that will meet the FDA's concerns for the particular drug. Prior experience in humans with evidence of safety may lead to reduced animal safety studies. Prior to Teva's licensing of copolymer-1 and assuming responsibility for development, the pilot human study discussed earlier was performed under an investigator IND submitted by Dr. Murray Bornstein, M.D., of Albert Einstein College of Medicine, supported by the preclinical studies just listed. The results of this placebo-controlled, double-blinded pilot study, completed in February 1985, were published in 1987 and will be discussed in detail in Chapters 14 and 15, but the fact that no serious adverse events were seen in the twenty-five patients who injected themselves with 20 mg of copolymer each day for two years provided evidence of safety for copolymer-1 at that dose. The adverse events seen more commonly in the patients receiving copolymer-1 than in those receiving the placebo included pain and redness at the site of injection and a transient feeling of difficulty breathing, flushing, heart palpitations, and anxiety immediately after injection. The breathing and heart symptoms lasted 5 to 15 minutes and resolved with no harmful effects. The reaction occurred three times in twenty-one months in one subject and twice in seventeen months in another. The medication was stopped in both patients (Bornstein et al. 1987).

Prior to submitting their IND and beginning the critical human trial in 1991, Teva performed a non-CGLP rat study of a single dose of 400 mg/kg (a thousand times the human dose by weight) injected beneath the skin, with no effects seen. A dog CGLP subchronic toxicology study in which doses of up to nearly one hundred times the human dose by weight were injected under the skin each day for four weeks was finished in 1988 after the results of the pilot human trial had been published. A number of CGLP animal safety and pharmacology studies were done to support Teva's development of copolymer-1. Subchronic studies in monkeys were done, dosing daily for a year with up to one hundred times the human dose by weight. Few significant toxic effects were seen in this monkey study beyond evidence upon postmortem examination of an inflammatory and antibody response in a few monkeys, which was not related to the dose given.

It was not possible to measure copolymer-1 in the blood, so for the animal pharmacokinetic (PK) and ADME studies, copolymer-1 was synthesized with one of its component amino acids containing a radioactive atom. The presence of radioactive material in blood was tracked in rats injected under the skin with the tagged copolymer-1. Blood was sampled frequently for eight hours after the injection of radioactive copolymer-1, and radioactivity was measured in the fluid part of the blood. To see whether repeated dosing changed the way the body handled the drug, this study was done in rats that had previously received no copolymer-1 and in rats that had received daily doses of copolymer-1 for 28 or 177 days. Similar methods were used in rat and monkey ADME studies, following the radioactive tag in blood, tissue, urine, and feces. This approach to PK is not ideal because the radioactive tag can change how the body handles peptides and because the method does not discriminate among the intact drug, a breakdown product of the drug, or a free radioisotope. Because no valid method was developed for measuring copolymer-1 in the blood, urine, and tissues, this was the limit of what could be done. The animal PK and ADME studies did support the idea that the drug was rapidly taken up and eliminated from the body (CDER 1996), and animal and subsequent human studies would support the safety of years of dosing with copolymer-1.

ANIMAL SAFETY AND PHARMACOLOGY STUDIES OF CELECOXIB

The animal safety, pharmacology, ADME, and PK studies performed to support approval of Celebrex (SC-58625, celecoxib) were much more extensive and more closely fit the FDA and ICH guidelines, as might be expected for a chemically more defined, smaller molecule, beginning development in 1995, for which the target population is essentially healthy individuals. More was also understood about the whole class of nonsteroidal anti-inflammatory drugs, from aspirin to ibuprofen, including the likely target for activity, possible unintended targets, and possible side effects (CDER 1998a).

The studies for the celecoxib IND included safety pharmacology, acute and subacute toxicology, and ADME studies. The chronic toxicology studies to assess the effects of long-term administration on the animals, their risk of developing tumors, and damage to their reproductive ability or to their offspring were completed while clinical trials were ongoing.

Researchers performed a laundry list of tests on mice to pick up changes in how the various parts of the body functioned after treatment with celecoxib. The list is long but does provide a sense of all the studies a sponsor does to support beginning human trials. To detect effects of the drug on the general functioning of the body, researchers treated mice with 50, 150, and 500 mg/kg of celecoxib. At the highest dose, the mice exhibited changes in general activity and behavior. Celecoxib had effects on drug-induced sleep as well as electroshock and chemically induced convulsions at doses above 150 mg/kg, indicating effects on the central nervous system. Detecting effects on pain again required researchers to cause pain in rats by either exposing their skin to concentrated acetic acid (vinegar) or by pinching their tail. Celecoxib decreased the response to vinegar but not to tail pinching. Celecoxib did not change the rats' body temperature at any dose. A test tracking the rate of passage of a charcoal meal in the small intestine picked up no effects of celecoxib, but the addition of high concentrations of the celecoxib to sections of small intestines of a guinea pig slowed the spontaneous movement of the intestine and reduced the muscle contractions caused by the addition of several agents known to trigger contractions. In dogs, doses of 200 mg/kg increased the rate of blood flow, but doses from 50 to 200 mg/kg brought no change in the electrocardiogram, blood pressure, or heart rate. Small changes in the volume of urine and its salt concentration were seen in rats.

In the acute toxicity studies, treatment of rats and dogs with a single oral dose of 1,000 mg/kg and 2,000 mg/kg led to no deaths or remarkable findings when the animals' organs were examined after the treatment. Oral absorption in dogs of a 5 mg/kg dose was greater if the drug was given with food. No monkeys given 25 or 250 mg/kg died. Watery or soft stools were the only finding, except for transient blood in the stool of one of three monkeys. The NOAEL single dose was 2,000 mg/kg in rats and dogs and <25 mg/kg in monkeys.

After a two-week feasibility study in mice of adding celecoxib to the food and several range-finding studies in mice and rats, rats received celecoxib for thirteen weeks at 20, 80, and 400 mg/kg by injection into the stomach. To provide information on the timing of the uptake of celecoxib from the stomach and its excretion, a group of the rats also received radioisotope-tagged celecoxib at three points during the thirteen weeks. Blood samples taken repeatedly during the twenty-four hours following the feeding of the tagged celecoxib indicated that the drug was well-absorbed, the peak blood concentration increasing with the amount given. These were important results for any new drug and for celecoxib specifically. Because the researchers believed that celecoxib would be safer than standard NSAIDs for the stomach lining, it was important to be sure that the drug is efficiently absorbed from the stomach. For any drug, it is important to know that in the range of doses that will be given to humans the amount of drug that enters the circulation is related to the dose given. Peak concentrations in the blood were higher in female than in male rats at the same dose. Only at the highest dose did repeated dosing decrease the amount of drug that ended up in the fluid part of the blood. The NOAEL in rats given celecoxib by mouth for thirteen weeks was 400 mg/kg.

To assess toxicity and reversibility of any toxic effects, dogs received celecoxib at doses between 25 and 250 mg/kg per day for four weeks. The study also included two subgroups of animals at 25 and 100 mg/kg that received radioactive drug on days one and fifteen to provide information on pharmacokinetics and metabolism. The intact drug and its breakdown products, detected and measured in blood, urine, and feces, provided information on how the body takes up the drug from the stomach into the circulation and how the drug is broken down and excreted. The concentration in the blood was related to the dose, and the blood level of the drug peaked at about two hours after dosing at the lower doses. However, the rate of clearance from the blood slowed after repeated dosing, suggesting the possibility that the drug might accumulate after repeated doses, at least in dogs. The NOAEL for four weeks of dosing was 25 mg/kg; dogs could not tolerate doses of 50, 100, or 250 mg/kg, primarily because of ulcers in the small intestine that led to bleeding and death. These data supported a starting dose in humans.

The subchronic study in dogs was a large and complex study that was designed to learn about the toxicity of the drug over thirteen weeks of dosing, about the reversibility of any toxic effects at the highest dose, and about how the drug was taken up, broken down, and excreted during the thirteen weeks. The dogs were divided into three groups given single daily doses of 15, 25, or 35 mg/kg, a group receiving 25 mg/kg in two divided doses each day, and a group receiving 35 mg/kg for the thirteen weeks followed by a twenty-eight day recovery period. One further group received isotope-tagged drug on day one and once during weeks six and thirteen. The major excretion was through the feces, as unchanged drug. Two major breakdown products were identified. There were no treatment-related changes in the dogs and at these doses no evidence that the absorption and clearance changed with repeated doses. Tests of preparations of the liver enzymes that carry out the conversion of the drug to its major metabolites showed no change in response to treatment. The NOAEL for dosing twice a day in dogs was 17.5 mg/kg.

To relate the concentration of celecoxib in the blood to damage to the stomach and intestines, researchers performed a seven-day dog study in which the drug was given directly into the vein. They measured blood and tissue levels of thromboxane, the product of the COX-1 enzyme, and prostaglandin $-E_2$ (PGE$_2$), the product of the COX-2 enzyme, to relate the drug level in the blood to these key mediators of tissue damage when inflammation occurs in target organs. Significantly, blood levels of thromboxane and PGE$_2$ dropped at both 15 and 40 mg/kg each day. High levels of PGE$_2$ were present in the stomach and colon but were little changed by treatment, which is consistent with a failure to inhibit COX-1. At 40 mg/kg, a dose equivalent to a human dose of 2.8 g, celecoxib caused ulcers in areas of the small intestine and evidence of bloody stool.

Daily doses of 20, 80, or 400 mg/kg to rats for twenty-six weeks resulted in death and gastrointestinal injury in two of twenty-five female rats at 80 mg/kg and six of twenty-five female rats at 400 mg/kg. No males died. The difference between the males and females may be related to the results presented earlier with radioactive drug and may be understood when we consider the three calculations made when drug levels are determined in pharmacokinetic studies. The C_{max} is the peak concentration of the drug found in the blood, T_{max} the time it takes to reach the highest concentration, and AUC the area under the curve when the blood concentration is plotted against time. The AUC reflects the total exposure to the drug with each dose. All of these parameters were 2–3 times greater in female rats than in males at every dose. Thus, the higher death rate in female rats may be tied to more exposure of the body's tissues and organs to celecoxib.

Two separate year-long studies in dogs, one with dosing twice a day and the other with dosing once a day, at doses of 0, 15, 25, and 35 mg/kg/day, showed neither significant treatment-related changes nor sex-related difference in pharmacokinetics. The dogs, independent of sex, did, however, fall into two groups, fast and slow metabolizers, based on a larger AUC, suggesting that individual dogs differ in their ability to break down celecoxib, though no link was made to toxic effects because since none were seen in these studies.

Long-term animal studies designed to see whether treatment with celecoxib increased the risk of cancer or had effects on reproduction were done as the clinical trials were ongoing. Two-year rat (0, 20, 80, and 400 mg/kg/day) and mouse (0, 25, 50, and 75 mg/kg/day for males and 0, 50, 100, and 150 mg/kg/day for females) studies showed no treatment-related increases in tumors. Gastrointestinal lesions were seen at all doses in male rats and at doses above 20 mg/kg/day for females. The NOAEL dose was thus 20 mg/kg/day for male rats but could not be established in this study for female rats because treatment-related lesions and deaths occurred at all doses. Tumors were not increased in the mice receiving celecoxib.

The reproductive study in which researchers dosed male and female rats before mating and females during pregnancy indicated a NOAEL of 600 mg/kg in males and 30 mg/kg for females in early pregnancy and 10 mg/kg for females in the latter portion of gestation. The NOAEL for reproductive toxicity in rabbits was 60 mg/kg. The NOAELs in all animal reproductive studies result in much higher blood and tissue exposure than that achieved by the human clinical doses of 200 and 400 mg per day (less than 6 mg/kg).

The gastrointestinal and kidney toxicity, and evidence of local infections in these tissues in dogs at doses at or above 50 mg/kg/day, signaled the need for close monitoring of human subjects for evidence of GI toxicity, bacterial infections, and kidney damage. The FDA reviewers in their analysis of all of the animal safety studies in the submission for approval for marketing celecoxib noted that the gastrointestinal tract and kidney were the main target organs for toxicity in the rats and mice. The reviewers also cited evidence that the drug entered, and may have caused tissue damage to, the brain.

Because COX-2 was found in human and animal nerve cells, they suggested that further studies of this issue would be beneficial.

RECALL THE PURPOSE OF THESE STUDIES

The fact that the animal studies resulted in toxicity is not in itself a concern. The animal safety and laboratory studies of drugs for humans are not intended to provide evidence of no toxicity. They provide an understanding of how the body takes up, distributes, breaks down, and eliminates the drug, what organs and tissues are at risk of damage, and at what level of drug in the blood and over what period of time of drug exposure the damage occurs. If the drug is only tested in animals at doses that do not cause toxicity, none of this will be known. In the beginning of the clinical trials, the animal safety studies provide clues about what might happen in humans, what is a safe starting dose, and whether the potential benefit of the drug offsets the risk of exposure of humans to the drug.

NOW TO THE TRUE TEST ANIMAL: WE KNOW WHAT DOSE TO BEGIN WITH AND WHAT TO LOOK FOR

Animal subchronic safety studies prior to the pilot human trial of copolymer-1 at 200–250 mg/kg in rats and 10 mg/kg in dogs showing essentially no toxic effects and the ADME study showing rapid elimination and no accumulation of the tagged drug allowed Bornstein et al. to start the 20-mg dose, equivalent on a weight basis to less than 0.5 mg/kg.

The acute (single dose), subacute (four week), and subchronic (thirteen week) studies of celecoxib indicated that in dogs doses up to 25 mg/kg given repeatedly did not have drug-related toxic effects. PK and ADME studies at these dose levels did not suggest that repeated dosing changed the rates of uptake and clearance or that the drug accumulated in the body. These results, plus the observed tissue damage at higher doses, thus provide the scientific basis for beginning human trials of celecoxib.

8

Getting Set for Clinical Trials

Before beginning clinical trials, a pharmaceutical company compiles and files an investigational new drug (IND) application (CDER, CBER 1995). If the FDA does not object, the IND permits shipping an unapproved drug across state lines for human use. The IND provides the documentation that the procedures and controls are in place to minimize the possible risk to human subjects and that the institutions and individuals responsible for carrying out the work are identified and acknowledge their responsibilities to carry out the studies in conformity with regulations (CDER, CBER 1995). The IND must include, in addition to the reports of the animal and laboratory pharmacology and safety studies described in Chapter 7, information about the drug, how it is manufactured, what is added to it in the pill or solution taken by the subjects, and what will be done to assure that the drug is provided at a consistent level of purity. The company responsible for manufacturing must be identified and its address provided so that the FDA knows it is an approved facility and can inspect it. The general plan for the studies and the detailed protocols for the initial, usually Phase 1, studies must be provided.

Just as there are CGLPs for laboratory studies, there are current good clinical practice (CGCP) guidelines for the performance of clinical trials for the registration of drugs for humans (ICH 1996). To provide a sense of how complex the process is, the CGCP guideline ends with a list of "essential documents for the conduct of a clinical trial" that must be developed and filed with the appropriate regulatory agency before the trial begins, during the trial, and when the trial is over. This list of documents numbers fifty-three, twenty to be generated before the trial begins, twenty-five during the trial, and eight when the trial is complete. Among the twenty to be submitted before the clinical phase begins are (1) the protocol signed by the investigator and sponsor, (2) the informed consent form including written information given to potential trial subjects, (3) the text of any advertisements to be used to recruit subjects, and (4) any financial agreements between the sponsor, the investigator, and the investigator's institution.

The CGCP focuses on the systems and procedures that must be in place for performance and documentation of clinical trials with a minimum of risk.

THE ETHICS OF HUMAN EXPERIMENTATION

Although good science is central to the clinical testing of a new drug in humans, the trials must be and must seem to be ethical (Emanuel et al. 2000). This has not always been true (Jonsen 1998, 2000). The modern era of concern for the ethics of human experimentation began at the end of World War II when the details of the horrific procedures performed on humans by the Nazis led to a conviction that such abuses should be prevented. Following the main Nuremberg trial by a military tribunal of judges from the Allied powers, the United States held tribunals for other elements of Nazi government, including the Doctors' Trial. Most of the defendants of the Doctors' Trial were physicians accused of murder and torture in experiments on concentration camp prisoners. The Doctors' Trial judges formulated a code of ethical principle, the Nuremberg Code, which provided rules for the protection of human subjects of research. In formulating the code, the judges acknowledged that the relationship between a researcher and subject was different from, but yet constrained by, the physician–patient relationship and the assumptions inherent in the Hippocratic oath and the pledge *primum non nocere*, above all do no harm. In the conventional physician–patient relationship, the patient comes to the all-knowing, beneficent healer and places trust in both the knowledge and ethics of the healer. When the patient becomes a subject in a research study, the physician is instructed by the Nuremberg Code not to let the demands of science defeat the pledge of the Hippocratic oath. The researcher is still a physician and the subject still a "patient," but the Code in its first principle changes the dynamic because the subject's voluntary and informed consent is essential (Nuremberg Military Tribunal 1949).

The Nuremberg Code

1. The voluntary consent of the human subject is absolutely essential.

This means that the person involved should have legal capacity to give consent; should be so situated as to be able to exercise free power of choice, without the intervention of any element of force, fraud, deceit, duress, overreaching, or other ulterior form of constraint or coercion; and should have sufficient knowledge and comprehension of the elements of the subject matter involved as to enable him to make an understanding and enlightened decision. This latter element requires that before the acceptance of an affirmative decision by the experimental subject there should be made known to him the nature, duration, and purpose of the experiment; the method and means by which it is to be conducted; all inconveniences and hazards reasonably to be expected; and the effects upon his health or person which may possibly come from his participation in the experiment. The duty and responsibility for ascertaining the quality of the consent rests upon each individual who initiates, directs or engages in the experiment. It is a personal duty and responsibility which may not be delegated to another with impunity.

2. The experiment should be such as to yield fruitful results for the good of society, unprocurable by other methods or means of study, and not random and unnecessary in nature.

3. The experiment should be so designed and based on the results of animal experimentation and a knowledge of the natural

history of the disease or other problem under study that the anticipated results will justify the performance of the experiment.

4. The experiment should be so conducted as to avoid all unnecessary physical and mental suffering and injury.

5. No experiment should be conducted where there is an a priori reason to believe that death or disabling injury will occur; except, perhaps, in those experiments where the experimental physicians also serve as subjects.

6. The degree of risk to be taken should never exceed that determined by the humanitarian importance of the problem to be solved by the experiment.

7. Proper preparations should be made and adequate facilities provided to protect the experimental subject against even remote possibilities of injury, disability, or death.

8. The experiment should be conducted only by scientifically qualified persons. The highest degree of skill and care should be required through all stages of the experiment of those who conduct or engage in the experiment.

9. During the course of the experiment the human subject should be at liberty to bring the experiment to an end if he has reached the physical or mental state where continuation of the experiment seems to him to be impossible.

10. During the course of the experiment the scientist in charge must be prepared to terminate the experiment at any stage, if he has probable cause to believe, in the exercise of the good faith, superior skill, and careful judgment required of him, that a continuation of the experiment is likely to result in injury, disability, or death to the experimental subject.

The issue of informed consent as the first principle of the code is striking, given that the transcript of the Doctors' Trial includes vigorous argument by the defense counsel that U.S. physicians in general and Dr. Andrew Ivy, M.D., the chief prosecution witness, had performed medical experiments on prisoners (Hornblum 1997; Shuster 1997). Ethicists understood that given their relatively powerless position, prisoners could not truly give voluntary consent. The examples of the U.S. use of prisoners for research dated back to cholera studies in 1906 performed by a U.S. physician in the Philippines on death row inmates (Hornblum 1997; Shuster 1997). Doctors' Trial defense attorneys also raised later studies on pellagra in a Mississippi prison, malaria in the Stateville Penitentiary in Illinois, and tuberculosis in a Colorado prison.

Doctor Ivy, a noted researcher and physician, then vice president of the University of Illinois and later infamous for his espousal of krebiozen, a questionable cancer treatment, in his testimony described the principles he had articulated for the AMA House of Delegates in 1946, providing for:

- voluntary informed consent,
- sound scientific foundation in animal experiments for the study "to eliminate any a priori reason to believe that death or disabling injury will occur,
- the absence of any other method to obtain the outcome, which must be useful to society, and
- the performance of the study by qualified individuals to avoid all unnecessary physical and mental suffering.

Ivy argued that prisoners, at least those in democratic countries, with the rule of law and respect for individual rights, could give voluntary consent (Shuster 1997). Adherence to these principles in the United States, particularly in the use of prisoners, was in fact poor during much of the twentieth century (Hornblum 1997; Lederer 1995). Current regulations define what studies involving prisoners are permitted and describe in detail the enhanced diligence in assuring that the rights of prisoners are protected and inducements are not coercive (CFR 2003a).

The United States and most of the world, faced with recurrent reports of unethical human research, have over the years built on and refined principles of the Nuremberg Code. The Declaration of Helsinki, adopted in 1964 by the World Medical Association and amended several times, incorporates much of the Nuremberg Code, but the Declaration distinguishes between patients participating in studies related to the diagnosis or treatment of their illness and healthy subjects who may be motivated by altruism and the promise of payment. Voluntary consent was viewed as more important in experiments involving healthy volunteers because the physician–patient relationship in trials involving the ill placed the physician in the conventional, paternal caregiver role, to which the patient is seen as ceding judgment (World Medical Association 2002a).

The international medical community supported the principles of the Declaration of Helsinki, but the Declaration did not have the force of law. In the United States, antivivisectionists, those opposed to the use of animals in research, had proposed federal legislation to govern research in humans as early as 1916, but it took over fifty years and several disturbing revelations to drive such legislation (Lederer 1995). In 1966, Henry Beecher, M.D., a Harvard anesthesiologist who had become an early proponent of placebo-controlled trials based on his observations of the demand or lack of demand for morphine by soldiers wounded in battle in World War II, published a report in the *New England Journal of Medicine* describing a number of federally funded experiments involving human subjects that ignored the rights of the participants to understand what was to be done and to give consent (Beecher 1966). He described studies in which retarded children were fed hepatitis virus, senile patients were injected with live cancer cells, thymus glands were removed from children undergoing heart surgery to see if this would have an effect on development, and nineteen other studies in which procedures of unknown risk were performed on subjects without informed consent. Beecher was direct in his outrage that these studies had gone on at prestigious institutions and that they had been federally funded.

In 1962, the Kefauver-Harris Amendments to the Food and Drug Act included limited statements about obtaining consent for subjects receiving investigational drugs, but Congress was reluctant to interfere with the physician–patient relationship, and the language of the amendments blunted the request for consent. As noted earlier, a major factor behind the development of the amendments was the disclosure that Richardson-Merrill Pharmaceuticals had distributed Kevedon (thalidomide) to academic physicians to test on their patients before the FDA had reviewed the application for approval. These physicians included obstetricians who gave the drug to their pregnant patients as a safe sleeping pill that was helpful for treating morning sickness. Doctor

Frances Kelsey, Ph.D., M.D., the medical officer assigned to review the application, had delayed approval of thalidomide based on her review of Richardson-Merrill's application even before she became aware of published reports in Britain and Europe of an epidemic of a particular kind of birth defect. Research early in her career had linked a specific drug side effect, peripheral neuritis, which in humans appears as tingling in the arms and legs and other symptoms, with birth defects in the offspring of pregnant animals treated with the same drug. Concerns about inadequacies in Richardson-Merrill's application plus the occurrence of peripheral neuritis in reports of English patients who had received thalidomide made her insistent on more safety studies in pregnant animals before she would sign off on approval of thalidomide. When the connection between thalidomide and the limb defects in infants was made, Richardson-Merrill withdrew its U.S. application for approval. Furthermore, the disclosure that over 200,000 women in the United States had received thalidomide, through Merrill's premarketing campaign, led to inclusion in the amendments to the Food and Drug Act the requirement that full and free consent be obtained from participants in clinical trials. Although the amendments did leave the need for disclosure to the best judgment of the physician, language requiring informed consent was included in regulations published by the FDA in 1966 (Harris 1964).

In 1966, the NIH, under the directorship of Dr. James Shannon, M.D., issued the Public Health Service Policy on the Protection of Human Subjects, which called for independent review of proposed studies involving human subjects to assure that the rights of the subjects would be protected and that voluntary informed consent would be obtained in written form (ACHRE 1997). The requirement for an independent peer review of proposed human experimentation was included in the NIH Policy as a result of a report prepared by a committee of the NIH chaired by the associate chief for program development, Robert B. Livingston, M.D. (Livingston 1964). The report noted an experiment performed at the Jewish Chronic Disease Hospital in Brooklyn in which elderly residents had been injected with cancer cells in an effort to study their ability to mount an immune response to the cells. Particularly disturbing was the willingness of both the medical director of the hospital and his collaborating researcher from

Sloan Kettering Cancer Institute to decide that the study was ethical. Livingston's committee concluded that the physician researcher could not be relied upon to protect the rights and interest of his patients and that an independent review must occur. The NIH Policy stipulated that institutions receiving federal funds had to provide the NIH with written assurance that they were complying with the requirements. The institutions were required to convene a local investigational review board (IRB) to review the proposed research protocols and determine whether the rights and welfare of the subjects will be protected, that the methods used to obtain consent are appropriate, and that the balance of risks and benefits of the study is acceptable. The onus was on the institution. The rules initially applied only to studies funded by the NIH at sites outside of the NIH but were eventually extended to clinical studies performed at the NIH (ACHRE 1997). The Declaration of Helsinki in its 1975 revision also called for review by a committee independent of the researcher and sponsor (the entity funding the study).

Another revelation of abuse of research subjects was the more widely publicized Tuskegee Syphilis Study. In July 1972, the *New York Times* reported on a forty-year study of syphilis carried out in Macon, Alabama, by the U.S. Public Health Service in which 600 black men, 399 with syphilis and 201 without, were followed from 1932 to 1972 (Reverby 2001; White 2000). The purpose at the outset had been to follow the natural history of the disease in blacks. The subjects were poor, uneducated rural men, seemingly recruited on that basis so that they would continue in the study in exchange for free medical exams, free meals, and burial insurance. They were told they were being "treated" for "bad blood," a term that described syphilis, fatigue, anemia, and other conditions. Although the study was planned to last only six months, the researchers worked hard and successfully to keep it going for decades. The news coverage of the Tuskegee Syphilis Study established it as a telling example of racism and ill treatment of African Americans. That Public Health Service and Veterans Affairs physicians had seemingly worked hard to continue the "natural history of syphilis" study, even after penicillin became known as an effective treatment for early syphilis and was widely used in that setting, is offered as evidence of the bias (Jonsen

2000; Reverby 2001). However, debates among the medical experts about the risks and benefits of treatment of late syphilis in nonsymptomatic patients raises the possibility that neither intent nor outcomes were so clearly unethical (White 2000). There seems to be little doubt, however, that the subjects were not asked to provide fully informed consent (Reverby 2001).

These exposes of the exploitation of poor, marginalized, defenseless subjects in studies by Public Health Service staff and faculty from prestigious medical schools that were funded by the NIH and other federal agencies roused both the public and the government. In 1974, President Nixon signed The National Research Act, creating the National Commission for the Protection of Human Subjects of Biomedical and Behavioral Research, which was charged with identifying the ethical principles that should underlie the conduct of biomedical and behavioral research involving human subjects and developing guidelines that should be followed to assure that such research is conducted in accordance with those principles. In 1979, the Commission published the Belmont Report, stating that in research with human subjects, three principles are central: respect for persons, beneficence, and justice (National Commission for the Protection of Human Subjects of Biomedical and Behavioral Research 1979). Respect for persons held that individuals should be treated as autonomous agents and that individuals with diminished autonomy should be protected. Beneficence requires that research should be carried out in a manner that maximizes possible benefits and minimizes possible harms. Justice requires that individuals should be treated fairly and equally, not depriving some to benefit others. Beyond setting out these principles to guide the design and performance of trials with human subjects, the Belmont Report indicated that in the application of these principles, researchers must provide subjects with comprehensible information so that they are able to freely and voluntarily give or refuse consent, that there must be a systematic analysis of the risks and benefits provided to the subjects and to society, and that this analysis should be reviewed by an independent committee. Thus, the focus of the Nuremberg Code on human rights is embodied in the Belmont Report as three principles—respect for persons, beneficence, and fairness—that are applied to how researchers should function. These rules have been incorporated into federal codes that describe the obligations of institutions that receive federal funds or companies that wish to have the FDA approve a new treatment (CFR 2003a; Office for Protection from Research Risks 1977). FDA approval of a drug is based on clinical trial data, and absent documentation that the trial was performed in accordance with the rules of the Belmont Report, codified in Title 21 of the Code of Federal Regulations, not only will approval be denied, but the FDA will refuse to review the data. Current ethical thinking holds that the science must be sound, the goal of the study must be valuable to society, and the subjects must be afforded dignity, justice, and beneficence. All these aspects must be independently reviewed by the IRB at each participating institution prior to beginning the study. Compliance with the rules for performance of clinical trials must be documented, and study sites are subject to inspections. Responsibility rests with each institution performing studies and with the physician investigators. Clearly, a sponsoring company is highly motivated to be certain that the rules are being followed and compliance documented. The cover page of the IND, above the signature of the sponsor or sponsor's authorized representative, states "I agree that an Institutional Review Board that complies with requirements set forth in 21CFR Part 56 will be responsible for initial and continuing review and approval of each of the studies in the proposed clinical investigation" (CFR 2003a; FDA 2002b). The consequences of failing to protect the rights of investigational subjects can lead, at their worst, to the loss of all federal funding to an institution, temporary or permanent debarment of an investigator from performing federally funded human studies or studies to be submitted to the FDA, and of course failure to gain approval or even review of an application to market a drug (CFR 2003a).

The regulations describe IRB makeup (five or more members with at least one member not associated with the institution), qualifications, frequency of meetings, processes for review, requirements of documentation of review, and exclusion of members with possible conflict of interest from review of a protocol. Each IRB at or associated with each institution, hospital clinic, contract clinical research organization, or doctor's office is essentially the gatekeeper for the conduct of human trials and considers not only the soundness of the science but also the protection of human subjects with regard to the principles of fairness, beneficence, and justice.

THE INVESTIGATOR'S BROCHURE

A large portion of the IND is taken up with the investigator's brochure (IB), a "compilation of the clinical and nonclinical data on the investigational products(s) that are relevant to the study of the product(s) in human subjects" (ICH 1996). The physicians and other medical personnel carrying out the study, those actually administering the drug, must have the information necessary to understand why the study is being done, why it is being done as described in the protocol, and what they need to know to provide responsible care for the subjects. The IB must contain results of all the relevant nonclinical studies of the drug and any previous experience with the drug in humans in sufficient detail so that a potential physician investigator can make an informed decision about whether to undertake the trial and how to care for the subjects, including what may happen to them as a result of administration of the drug. Although the IB is first prepared for the IND as clinical trials begin, it must be updated regularly and reviewed at least once a year.

The IND must also include a general plan for the investigation and the protocol for the initial clinical trial. A clinical protocol provides the description of the purpose, plan, and procedures for the study. Who will receive the drug, who will be excluded for safety concerns and how many subjects will be tested, and the plan for how much drug will be given and for how long must be stated. The monitoring of subjects for adverse events must also be described (CDER, CBER 1995).

SOMETIMES NO NEWS IS GOOD NEWS

The FDA has thirty days to review the IND and decide whether the trial should be halted. If no notification of a clinical hold is received within thirty days after the FDA has received the IND (and this should be documented because documents can get lost), the sponsor may begin clinical testing (Mathieu 1994).

Any changes in protocols must be communicated promptly to the FDA, particularly amendments that may have an impact on the safety of study participants. Review by the IRB and approval of changes must be documented (Mathieu 1994). Annual reports are also required. But with an IND filed and no clinical hold imposed, the trials may begin.

9

Phase 1 Clinical Trials

Consider Mary, who has chosen to participate in a Phase 1 clinical trial. After seeing an advertisement posted on a grocery store bulletin board, she called the phone number and, after answering a few questions about her age, height, weight, and general health, she was given an appointment to visit the testing facility, learn about the study, and have some blood taken for lab tests. The facility was clean; everything looked neat and sort of new, almost like a regular office. She was told that the study was for a new medicine to treat arthritis and required that she stay at the testing facility for 7 days. She answered more questions, and they took several tubes of blood and said they would call her when the tests were done. A week or two later, they called Mary to tell her all her lab tests were good and she should come in for a physical exam from a doctor. Before her exam, they gave her many pages to read that described the study and what would happen to her, and what the side effects of the drug might be. She watched a video that explained more about the study and met with a staff member who answered her questions. Then she signed the informed consent document stating she understood what she had read and agreeing to participate in the study. A doctor gave her a physical exam and an electrocardiogram. Although she had signed the informed consent form, during the exam the doctor quizzed her about the study to make sure she understood what she had read. That was after he asked her, and this was the third time they had asked, whether she was taking birth control pills. They seemed to be most worried that she was pregnant, though the pregnancy test had been negative. They said the test could still be negative if she was only a few weeks pregnant. Mary knew she was not pregnant.

Before her exam, they had shown Mary where she would stay, the place they called the clinic. It was a big room filled with rows of comfortable-looking chairs, windows to a courtyard with trees, plants, and benches on one side, a small room on the other side where she would be given the drugs and have her blood drawn, a giant TV, and a pool table. Maybe she would learn to play pool, the person giving her the tour had said. Off in the corner were the ten or so small cubicles, each about as big as a department store dressing room, with two bunk beds in them. That was where she and the other women would sleep. She hoped that her roommate did not snore.

Checking in on the first day, Mary was surprised to see so many people in the clinic. She knew that there were only twenty women in the study, but the clinic just buzzed with technicians, nurses, aides, or what they called coordinators. When Mary heard them called that, her first thought was that their assignment was to make her less clumsy; but she knew that was silly. These were the people getting them settled in, helping them put away their stuff. They seemed as nosy as helpful, but Mary knew they had to check to make sure she did not have any pills or cigarettes, or even LifeSavers. They had said the study was called a Phase 1 study, and one thing they wanted to find out was whether the drug got to your blood faster if you took it on any empty stomach or after you had eaten, so sucking on a candy could mess up the results.

One thing Mary worried about was all the times they had to take a blood sample. She was not a baby about needles, and the nurses never had a problem getting a blood sample when she went to the doctor, but getting stuck eight times after the first and last pill was scary. Well, it was now her turn to get the hospital bracelet that they used to make sure they had the right person for every pill and every blood sample. As if she would lie!

PHASE 1 TRIALS

The purpose of the Phase 1 trials, often the first introduction of the drug into humans, is to provide assurance of the safety of the drug. This is done by administering varying amounts of drugs to human volunteers and monitoring them for any adverse, toxic effects. In Phase 1, how the human body absorbs, distributes, and eliminates the drug will also be tracked. Phase 1 participants are monitored for changes in vital signs, pulse, breathing rate, and blood pressure. Adverse events, changes in how the subject feels, or changes in the vital signs are recorded. Blood, urine, and stool samples are taken and tested for levels of the drug and its breakdown products. The blood levels of the drug and its breakdown products will be related to other laboratory tests and the clinical measurements, such as blood pressure, heart rate, temperature, and so forth depending on the nature of the drug and the results of nonclinical pharmacology. Phase 1 participants may be given a form of the drug tagged with a radioactive molecule to more accurately track how the body distributes and eliminates the drug. The detailed plan for the trial—the characteristics for who will be included in the study, who will be excluded, how and when the drug will be given, what lab tests and procedures will be done, what information will be collected, and how it will be analyzed—is called the protocol.

For many drugs, Phase 1 studies are done in healthy adult volunteers in an inpatient medical unit specifically set up for the purpose with the necessary medical, nursing, and laboratory facilities. Such Phase 1 testing facilities may be within hospitals or research universities but often are free-standing commercial facilities. Phase 1 trials of certain drugs, such as cancer or AIDS drugs or drugs that suppress the immune system, because of their risk, are generally not done in healthy volunteers but in people with the condition for which the drug is intended. Although the primary focus of Phase 1 trials, whether in healthy volunteers or patients, is safety, evidence for activity against the targeted condition may also be sought.

FIRST HUMAN STUDIES OF COPOLYMER-1

The initial human tests of copolymer-1 do not fit the current standard Phase 1 scenario, in part because the drug was discovered and first tested outside of the United States many years before the guidelines for such studies were developed (CDER 1996a). Also, the absence of harmful effects in animals even at very high doses and the grim prospect of MS patients at the time supported an unconventional approach. The first human tests of copolymer-1, a small Israeli trial in three patients with an acute brain inflammation and four patients with advanced MS, resulted in no adverse side effects (Abramsky et al. 1977). The brain inflammation patients recovered, though the authors state that about half of similar patients treated with steroids also recover. The MS patients did not improve. But no side effects were observed in any of the seven people treated with copolymer-1. The brain inflammation patients received 2 mg of copolymer-1 daily for 2 weeks. The MS patients were given 2–3 mg every 2–3 days for 3 weeks while they were in the hospital, and then 2–3 mg each week for 2–5 months.

PHASE 1 TRIALS OF CELECOXIB

The initial clinical trials of SC-58625, celecoxib, are consistent with guideline documents published by the FDA, specifically the Guideline for the Clinical Evaluation of Analgesic Drugs and the Guidelines for the Clinical Evaluation of Anti-Inflammatory and Antirheumatic drugs (CDER 1992, 1977). The first Phase 1 study of SC-58625 was designed to determine the safety and tolerability of a single oral dose of the drug and find out how human males took up and eliminated the drug. The dosing started at 5 mg. Remember that an average male weighs 60–70 kg, so we are talking about less than 0.1 mg/kg, and the NOAEL (no observed adverse effect level) in dogs for multiple doses was 25 mg/kg. Why the big difference? The Phase 1 starting dose, particularly when the study is the first time the drug is being given to a human, is calculated to provide a wide safety margin, correcting for differences in the size between specific animal species and humans. Body surface area has become accepted as the most important way to express size for drugs, based on studies of the toxicity of cancer chemotherapy drugs (CDER 2002a). The dose in humans that is equivalent to 25 mg/kg in dogs using an established correction for the surface area to weight difference between dogs and humans is

13.5 mg/kg; putting in the usual safety factor of 1/10 results in a starting dose of 1.35 mg/kg, or 94.5 mg. But the 1/10 safety factor may need to be decreased for a number of reasons, including if the toxic effects are difficult to detect, severe, or come on without warning. The key toxicity of SC-58625 in animal studies was damage to the lining of the stomach and intestines, which is certainly difficult to detect. Also, the whole rationale for a selective COX-2 inhibitor is effective anti-inflammatory activity without toxicity. The goal, unlike the situation with conventional cancer chemotherapy drugs, is not to give as much drug as possible but to achieve a blood concentration for a reasonable period of time that causes inhibition of COX-2 activity but does not inhibit COX-1. The concentration of SC-58625 in the test tube needed to inhibit COX-2 to half was about 0.04 µg/ml; inhibiting COX-1 required 10–20 times as much. Therefore, a reasonable approach to human testing was to start at a very low dose and increase the amount, measuring the maximal concentration of the drug achieved at each dose as well as the time it took to reach the peak concentration and clear the drug from the blood. After fasting overnight, Phase 1 subjects took a pill of 5, 25, 50, 100, 200, 400, 600, 900, or 1,200 mg, and blood samples were taken at precise times just before and multiple times after dosing. Peak concentrations of the drug in the blood were reached at two hours for all levels of dosage. The total exposure to the drug is captured and visualized by graphing the blood concentration over time. The area under the line connecting the points is called the area under the curve (AUC). For the single-dose celecoxib Phase 1 study, the AUC was proportional to the dose up to 600 mg, providing confidence that the amount of drug available to the body was predictable and dependent on how much was swallowed. A few subjects at the 200 and 400 mg levels were given a second dose after a high-fat breakfast.

The next study was a repeat dosing study done in healthy adults with blood level testing. A single dose of 40, 200, or 400 mg was taken and then two days later each subject took the same dose twice a day for seven days, with regular blood sampling after each dose. This study provided important information beyond evidence of safety, showing how the body absorbed and eliminated the drug with twice-daily dosing. This information showed that the body's exposure to the drug stayed more or less the same with multiple days of drug and was proportional to the dose. These pharmacokinetic studies allow the selection of doses and schedules for further trials for efficacy, doses that were not only safe but would provide a relatively steady concentration of the drug during the entire treatment period.

But a few more studies were needed for planning the efficacy trials in patients. The effects of taking the drug while fasting or after eating were tested, as were the effects of taking an antacid with the SC-58625. Because people who would be taking the drug, people with osteoarthritis, were likely to be older than the 18–44 years of the initial single- and multiple-dose studies, a study of how the drug at 40, 200, or 400 mg twice each day for fourteen days was handled by the body in a group of 40–58-year-olds was done. The clearance was somewhat slower in this age group, and thus total exposure of the body to the drug was higher with each dose. Another study, which compared the timing of the appearance and disappearance of the drug in the blood of a 200 mg dose twice a day in twenty-four healthy individuals less than fifty years old with twenty-four elderly (≥65 years) individuals, was done and confirmed the 60–80 percent increase in drug exposure in the older individuals. The differences were greater for elderly females than males. Subsequently, Phase 1 studies were done in healthy elderly individuals with kidney function somewhat less than normal and in patients with mild or moderate reduction in liver function. How the drug was handled in the body for the last group was important because SC-58625, like so many drugs, is broken down in the liver, and decreased liver function would mean the drug would stay in the body longer at higher concentrations, perhaps causing tissue damage. For the same reasons, pharmacokinetic studies were done in subjects taking other drugs to see the impact of taking SC-58625 on how the body handled methotrexate, a drug often used in rheumatoid arthritis, lithium, tolbutamide, a drug used in adult-type diabetes, and warfarin, a blood-thinning drug. These studies were done as the human trials for efficacy proceeded so that information was available to allow the safe use of SC-58625 in people with more than one illness who were taking other drugs.

To find out how the body takes up and metabolizes SC-58625, a small number of healthy males received a single dose of 300 mg

SC-58625 synthesized to contain ^{14}C, a radioactive isotope. At timed intervals, radioactivity was measured in the plasma, the fluid part of the blood, the red blood cells, saliva, urine, and stool. The maximum radioactivity in the plasma was at 1¾ hours and was undetectable by seventy-two hours. Red cells did not have a higher amount of radioactivity, indicating they were not picking up and concentrating the drug. The amount in saliva was negligible at all time points. Excretion was 27 percent in the urine and 58 percent in the stool. Most of the radioactivity was excreted in the stool and urine by ninety-six hours. These results provide confidence that SC-58625 would not accumulate in the body. The major chemical form of the drug in the plasma was SC-58625 itself, indicating it would not be broken down extensively before it could reach the site of inflammation. Three breakdown products, metabolites, were identified in the plasma at three, four, and twelve hours after dosing. No unchanged drug was found in the urine; two of the three metabolites were found in the urine. Radioactivity in the stool consisted mostly of one metabolite and a very small amount of unchanged SC-58625. Based on these results, the pathway of breakdown of SC-58625 by the liver, where most chemical breakdown of drugs occurs, was determined and confirmed by studies in the laboratory of the breakdown by liver extracts and liver enzymes.

No adverse events were reported in the Phase 1 studies of SC-58625, and thus the stage was set for tests of whether SC-58625 was safe and effective in settings where COX-2 activity was thought to contribute to symptoms, including pain after surgery and both rheumatoid arthritis and osteoarthritis.

10

Phase 2 Clinical Testing

What was the situation facing someone who was diagnosed with MS in, say, 1985? Lots of tests, lots of people telling you to hope, but no treatment that would hold off a future that might include a wheel-chair. This was the future facing someone asked to volunteer for a trial of a new drug that had cured mice, guinea pigs, and monkeys but had little or no track record in people with MS, except (as determined in the Phase 1 trials) that it seemed to be safe.

Consider Susan, a 33-year-old mother of two and finance direc-tor for a real estate developer. It was Monday morning, and she had to get up and make breakfast and pack lunches. She had to put on her suit, pack her briefcase, put the kids in the car and drop them at daycare, and get to work. She had that report on the financial plan for the new shopping center due. It was nearly done; she just needed to get a firm quote from the landscape management company. But she could not do any of these things. She realized that when she opened her eyes, before the alarm went off as usual. She could not do any of these things because her left leg was numb and would not move. She did not think she could even get to the bathroom without falling, and she needed to get to the bathroom. If she wet the bed, Jeff would wake up and figure this out, but she really did not want to wake him that way. So, she stretched out her right arm (at least that worked) and rubbed his arm, and said "Jeff, Jeff, honey I need some help."

This was not the first time this, or something like this, had happened. The first time, about five months ago, she had not gently rubbed her husband's arm; she just yelled. Not knowing what was going on, they decided he should take her to the doctor. He had helped her get up, sort of, helped her to the toilet, bless him, called his Mom to come over to take care of the kids, and then called their doctor, Ellen Jones. Doctor Jones said they should go to the emergency room. Jeff then carried and dragged her to the car, drove to the hospital downtown, and got a wheelchair to take her into the ER. The hospital staff took over in the ER, but only after a 45-minute wait. Susan was glad to see Dr. Jones when she arrived. After a brief, whispered conversation with Dr. Jones, the emergency room doctor examined Susan and then did some weird stuff. He asked if she could see his wiggling fingers when they were on one side or the other of his outstretched arms, made her stick out her tongue, touch her nose with each hand with her eyes shut, and checked whether she could feel pin pricks on her legs, arms, belly, chest, and face. The ER doctor asked whether this had happened before to her or to anyone else in her family. She said no, and then he went away to call Dr. Jones, who by then had left to start her office hours. Both doctors agreed that a neurologist should be called in and more tests done. They were not sure what was going on but suspected Susan was having an attack on the insulation of some nerves. They admitted her to the hospital for what turned out to be five days, gave her large doses of steroids through a vein to calm down the attack, and they ordered some tests. She had a spinal tap, which did not scare her because she had had an epidural when Sally, their second baby, was born. But the next test, the MRI, did scare her. It didn't hurt; it just was strange. You lay down on this sort of bed, after taking off your jewelry, and they slide you, head first, into a machine, into a tunnel really, and then take some kind of pictures while you hear this dull drumming. It took so long, though it was done in stretches and they would talk to you through a microphone while you were in the tunnel, saying the next set would take two minutes or five minutes or whatever. Those minutes went very slowly. Then they slid you out of the tunnel, injected something, gada-something, they said, into a vein in your arm and did the whole thing again.

The next day, the neurologist that Dr. Jones and the resident had called to consult saw Susan, and he laid it out for her. Her leg was not working because her body's immune system was attacking the insulation on the nerves from her brain and spinal cord that controlled the leg. The steroids would help her get over the attack, and she would probably get back to normal in a few days. Her leg was already beginning to work better; she had taken some steps that morning. They would stop the steroids in a few days and she could go home and be back at work in a week or so. The problem was that they did not know if it would happen again. More than half of people who had an attack like this never had another, but some did, which meant it was multiple sclerosis. She did not know what that meant. Was that what Jerry Lewis had all those telethons for? Wasn't that a kids' disease? No, they said this was something else, a disease that usually hit women in their 30s like her. But they told her not to worry; the chances were good this would not happen again. Don't worry! Susan started to cry; Jeff, who told her later he had suspected something like this, just clenched his jaw.

Now here she was, having another attack. What was going to happen to her? They told her there was no cure and no treatment to stop the attacks. They said there was a lot of research going on and that everyone was very hopeful. In fact, the hospital was running a study for a drug that might help. But right now, none of that helped her. Susan still had to get up and make those sandwiches. If she didn't, who would? So now with two attacks and the MRI evidence, she had MS and would be asked whether she was interested in enrolling in a trial for a new drug that might stop the attacks.

PHASE 2 TRIALS

Phase 1 trials establish the safe limits for dosing the compound in healthy people and provide some understanding of how the human body handles and responds to the compound. But is the compound useful? In the industry shorthand, is it a drug? That is what Phase 2 trials are for—to find out whether the compound might be useful (effective and appropriately safe) in one or more medical conditions. Following careful review of the data from the Phase 1 trials, the sponsor will begin Phase 2 trials, in subjects who have a condition for which the drug is intended, to develop evidence of effectiveness of the drug for the intended use. Phase 2 trials also provide short-term evidence of safety in this population. In Phase 2, the amount of drug given, the way it is given (for example, by mouth or by injection), and how many times and how frequently the drug is given will often be tested to establish the best way of administering the drug.

Phase 2 trials, while maintaining focus on the safety of the compound, begin the effort to provide evidence of efficacy. Just as the animal studies earlier provided proof of concept, these studies provide proof of concept in the intended users, human beings with a specific medical condition. Small early Phase 2 studies often provide the sponsor with the information for a go/no-go decision, which is critical because the larger, usually longer, Phase 3 trials, which are pivotal for approval, demand a greater investment in time, money, and human resources. The sooner managers of a pharmaceutical company can find out that a drug is highly likely to fail, or at least not be a big enough success to gain approval and sell, the more successful the company will be in the long run, if the company has other drugs that seem more promising. Small, new companies with a single compound ready for clinical trials are often gambling the survival of the company on Phase 2 trials.

Phase 1 trials, particularly those performed on healthy volunteers, are often carried out in dedicated Phase 1 facilities run by the pharmaceutical company or by contract research organizations (CROs). Phase 2 trials, because they are performed in patients, are usually carried out in physicians' offices, hospitals, or university research centers staffed by nurses, technicians, and study managers. At each study site (a large study may involve many sites), a physician is responsible for the study, and the institutional IRB must review and approve the study protocol. The performance of clinical trials in hospitals and university research centers has developed into a very businesslike undertaking (Anderson 2003). The company sets up a contract with the site and pays the site for the costs of carrying out the study, based on the personnel required to administer the drug and perform any procedures and laboratory tests required by the study. Procedures, tests, and drugs that would have been given to the patient for his or her routine care are generally billed to the patient's health insurance provider. The company must also pay for the legally required initial and annual review of the protocol by the IRB and the collection and management of the data, the information collected about each subject before, during, and after the treatment period. The contract will also cover the costs for any advertising and recruitment of subjects. Professional services by physicians and other health care workers beyond that required for the routine care of the patients are also covered. Companies, as part of CGCP, will send clinical monitors to the site to review the study. This is done specifically to compare each patient's chart with the case report form (CRF), the document into which all the study-related data are entered (with the patients' names replaced with a number), to make sure that the data are being entered accurately. Each such visit creates work for the site-data coordinator, and that generates a charge. If the protocol is changed, then additional costs for the IRB review of the changes will be charged. The list goes on. Costs vary with the size, duration, and complexity of the study but studies can cost the sponsoring company $2,000–$10,000 per subject or more.

While the clinical trials are going on, the investigators are required to report any serious adverse events in a subject to the sponsor and to state whether the adverse event was likely or unlikely to have been caused by the treatment. A serious adverse event is defined as any untoward medical occurrence that at any dose results in death, is life-threatening, requires inpatient hospitalization or prolongation of existing hospitalization, results in persistent or significant disability or incapacity, or is a congenital anomaly or birth defect (ICH 1994).

If the adverse event is both serious and unexpected, the FDA must be informed quickly, so that other investigators can be notified and decisions made to change the dose, modify the trial, or simply halt the trial. Reporting an unexpected SAE may require breaking the blind, disclosing to the investigator and sponsor whether the patient received the experimental drug or not. All SAEs must be reported to the FDA and included in regular safety reports to the FDA or another appropriate regulatory agency (ICH 1996).

All of the adverse events must be recorded during the clinical trial and any follow-up period. The adverse events are described and also coded using a mapping dictionary that provides codes for symptoms and changes in laboratory tests, and organizes them into relationship groups based on the organ or tissue involved. The currently accepted mapping dictionary is called the Medical Dictionary for Regulatory Activities (MedDRA), developed by a committee of the ICH and adopted by the FDA as well as Japanese and European regulatory authorities (ICH 2003a). This allows consistency in reporting adverse events and the ability to see patterns quickly and unambiguously. For example, if the staff in one hospital records that a subject is not eating and another describes the same situation as anorexia, MedDRA software suggests that both be coded and recorded as anorexia.

Phase 2 studies are generally controlled, randomized, and double-blind. The scientifically rigorous tests of whether the experimental drug has beneficial activity and safety in patients are designed to provide statistically robust tests of the null hypothesis, the hypothesis that the new agent is *not* more effective than the control. For example, the null hypothesis for a trial comparing celecoxib to ibuprofen for pain after having a tooth pulled is that celecoxib and ibuprofen provide the same level of pain relief. The criteria on which the exclusion of the null hypothesis are based, called endpoints, are characteristics of the medical condition being studied, as recognized and accepted by medical experts in the field. Phase 2 endpoints may include laboratory tests or other measurements that are accepted surrogates for medical improvement of the subjects, but medically important endpoints, such as a lower frequency of death, and a higher frequency or extent of recovery, are very important. After all the studies are completed, the FDA approves the drug as safe and effective. The criteria for efficacy must include a benefit apparent to the person treated; improvement in a surrogate endpoint may or may not necessarily translate into a benefit to the patient. The multiple endpoints will often be ranked as primary, secondary, and so on based on the extent to which they reflect medical benefit and their importance to the goals of the trial. Phase 2 trials usually involve up to several hundred patients and may take several months to two years to complete (Mathieu 1994).

Reliance on randomized, controlled, and blinded trials to decide whether a medical procedure or drug is safe and effective has evolved as medicine has become more science than art and as appreciation has grown for the risk of drawing conclusions too easily (Fisher 1951). Clinical trials are essentially the application to medicine of the fundamental scientific method of hypothesis testing, so that physicians, relying on the results of a study on a sample of patients, can make treatment choices for a much larger group of people. The physician and the patient base their decisions on the results from a small sample of people, and the methods of statistics are used to provide confidence in predictions based on the sample. People vary in their response to any medicine—some benefit, some do not, some experience severe side effects, others experience few or mild side effects. Genetics and many other factors control this variation. Take the question of 200 people taking a dose of a drug being tested to speed the healing of a surgical wound or placebo. We will score the extent of healing on day ten on a scale of 1 to 10, where 1 is no improvement and 10 is a wound that is completely healed. If you make a graph of how many people within each group score 1, 2, 3, 4, 5, and so on, the scores for each group will cluster around a number. Let us say that most people taking the test drug score an 8. Just because of the variation inherent in being human, some will score 10, some 9, and so on for each score even down to 2 and 1. The placebo group might fare much differently, with most people scoring a 4 and a few 5, 6, or 7. Based solely on the mathematics of chance and error, the graphs of each group will likely look like the infamous bell curve, with the peak at 4 for the placebo and at 8 for the test drug. The two curves, if plotted on the same graph, may overlap a bit. The null hypothesis is excluded if there is a high level of confidence that the two curves are different not just by chance because only a sample of people were enrolled in the study. Statistics allow the researchers to put a number to their level of confidence such that if a very large number of people used the test drug under the same conditions, most people would experience wound healing and would score higher than those given the placebo. So though the causes for variation in response, good or bad, may be poorly defined, statistics are used to provide a level of confidence that a patient subsequently treated with the agent being tested has a defined and reasonable likelihood of benefiting from the treatment.

Sham, or placebo, controls are used to reduce the possibility that suggestion will influence the outcome of a test of a new drug. If the control subjects are not going through the same procedures, taking pills or injections, then blinding is lost. If the controls are subjected to the same procedures as those receiving the test material, then the result of the comparison is more likely to have been caused by the material itself and not self-suggestion or conscious or unconscious bias on the part of the physicians or nurses doing the assessments. Although the placebo is an inactive substance (a sugar pill or saline injection), there is a widely held belief that there is a placebo effect—that subjects or patients may well experience some improvement having received the placebo, perhaps because of the psychological effect of wishing or believing that what was taken was the active drug. This is not only an issue in clinical trials: Physicians have long debated the practice of giving a worrisome patient a prescription for sugar pills without disclosing the actual nature of the prescription, based on the idea that the suggestion of effect would itself bring about a beneficial effect bolstered by the psychological effect of the physician taking any action, relevant or not. Although such practice tends to be frowned upon in modern, mechanism-based medicine, with the focus on patient autonomy, physicians may well hold to the idea that their intervention has in itself great power. Hubris aside, research has suggested that there is little sound evidence to support a placebo effect. In fact, a thorough analysis of published studies that included both a placebo and a nonintervention arm did not support the existence of a discernable placebo effect (Bailar 2001; Hróbjartsson and Gøtzsche 2001). In clinical trials, however, the use of a placebo control arm, particularly when there is no active control arm, is the standard, for no other reason than to sustain blinding. Placebo controls are not used where the condition under study is serious and an approved agent that can alter the course of the disease is the standard of care. To deprive a participant of the standard of care or the test agent under study to see whether the test drug is effective is unethical (Levine 1988). Investigators, as physicians, also face ethical concerns in enrolling a patient in a blinded, controlled study for a new drug against an effective, approved control treatment. How can they gamble with the subject's health or life and risk their being randomized to receive what turns out to be an inferior treatment? The resolution to this ethical problem is hotly debated but usually comes down to the question of clinical equipoise, the fact that the medical experts themselves are uncertain whether one treatment or the other is better (Freedman 1987).

PHASE 2 TRIAL OF COPOLYMER-1

The Phase 2 study of copolymer-1 was a randomized, placebo-controlled, double-blind trial in fifty patients with the exacerbating-remitting form of multiple sclerosis (Bornstein et al. 1980; Bornstein et al. 1987). The 85 percent of MS patients who have this form of the disease experience temporary significant worsening of symptoms one or more times a year. The sudden worsening of symptoms or onset of new symptoms, called an attack, a relapse, or an exacerbation, can involve the eyes, causing double vision or loss of vision in one or both eyes, or significant weakening of one or both legs or arms to the point of paralysis. Generally, the symptoms will resolve more or less over a period of weeks with no treatment, though the general practice is to treat the exacerbation with a short course of corticosteroids, administered into a vein to reduce the inflammation and help speed resolution of the symptoms. The patients may return to baseline after an attack, or some of the symptoms may not resolve completely. This accumulation of neurological dysfunction is tracked by one of several scoring symptoms. An increase in the score is evidence of progression of disease; the higher the score, the more disabled. The scoring system in use, the EDSS (expanded disability scoring system), ranges from 1 (no disability) to 10 (death). The fifty subjects enrolled in the trial, drawn from a group of over 900 people, had to be between twenty and thirty-five years old, have experienced at least two documented exacerbations in the preceding two years, be emotionally stable, and have an EDSS score of 6 or lower, meaning they are able to walk with assistance. The fifty subjects injected themselves daily for two years with either 20 mg of copolymer-1 in saline or saline alone. Each subject was seen and evaluated by a neurologist, who was unaware of the patient's group, at the beginning of the trial, one month after the trial began, and then every three months for two years. The neurologist would assess how the patient was doing, particularly any change in disability. Patients were also seen when they suspected they were having an exacerbation to provide objective documentation of an exacerbation.

The primary endpoint of the study, the way the two groups would be compared, was the proportion of patients with no exacerbations. The other endpoints were the frequency of exacerbations, the change in the disability score, and the amount of time until a measurable increase in disability.

Two patients in the placebo group were dropped from the trial for what were said to be psychological reasons, so the results of forty-eight patients were analyzed. The results for the primary endpoint were clear and statistically significant: Fourteen of the twenty-five patients in the copolymer-1 group and six of twenty-three in the placebo group had no exacerbations during the two years of the study. There were 62 exacerbations in the placebo group of 23 and 16 in the copolymer-1 group of 25, resulting in two-year average rates of 2.7 in the control group and 0.6 in the copolymer-1 group. The effects on disability were not as clear. The less disabled patients taking copolymer-1 improved slightly in their disability scores, whereas the less disabled taking the placebo worsened. More disabled patients in both groups worsened slightly.

The investigators stated that the results "suggest that Cop 1 (copolymer 1) may be beneficial in patients with the exacerbating-remitting form of multiple sclerosis" but caution that the results are preliminary and will need confirmation in a more extensive trial (Bornstein et al. 1987).

Local side effects at the site of injection of soreness, itching, and swelling were more common in the copolymer-1–treated patients. Medically more significant was a complex of symptoms that occurred rarely right after injection of copolymer-1. The symptoms involve a feeling of tightness in the throat, a red face, and heart palpitations. Doctor Bornstein called it a vasomotor event because it appeared to involve an increase in the diameter of blood vessels (thus the red face and heart palpitations). It would later be called, for want of a better term, the systemic reaction because it seemed to involve the whole body. This rare event would continue to be a feature of the use of copolymer-1.

PHASE 2 TRIALS OF CELECOXIB (SC-58635)

At the outset, three indications were targeted for SC-58635—postsurgery pain, osteoarthritis, and rheumatoid arthritis—three settings where inhibition of cyclooxygenase might help and where clinical trials might be straightforward. The first Phase 2 study of SC-58635 was a single-dose, single-blind comparison of SC-58635 at 100 mg (fifty patients) or 400 mg (fifty patients) to a placebo (fifty patients) or 650 mg of aspirin (fifty patients) in patients who had undergone dental surgery, specifically the removal of one or more third molars. One goal was to see whether either dose of SC-58635 provided pain relief superior to a placebo (yes) or aspirin (no), based on whether the subjects required further pain medication. This was the first test for the ability of SC-58635 taken once to reduce pain in humans, a critical first step in the development of a pain reliever, according to the FDA guidelines (CDER 1992). The guidelines for pain relievers suggest such a single-dose study be performed before starting a multidose pharmacokinetic study in healthy subjects so that there would be some level of confidence that the test drug might be able to relieve pain. The idea is to provide a quick, simple test that can weed out drugs that have no future. Although more multidose postsurgery, dental, and orthopedic studies would be done, the results of this initial single-dose study would set the stage for all the pain studies.

The first Phase 2 study in osteoarthritis was a two-week, randomized, double-blind study in the United States that compared SC-58635 at 40, 100, or 200 mg taken twice a day with a placebo in individuals with osteoarthritis. A dose was found that did improve the symptoms of arthritis over the placebo (Simon et al. 1998a). A subsequent 402-patient dose-ranging (25, 100, or 400 mg taken twice a day) placebo-controlled four-week study in osteoarthritis demonstrated that 25 mg was ineffective, using for the primary endpoint the answers to four questionnaires about their arthritis and pain that were given to the patients and their physicians. A dose of 100 mg of

SC-58635 was superior to the placebo based on the patient's and physician's overall sense of the effects on the arthritis, but not with the other questionnaires used to ask specifically about the effects on pain or the severity of the arthritis; 400 mg taken twice a day was better than placebo based on all the endpoint tests.

A Phase 2 study in rheumatoid arthritis, a form of arthritis in which joint inflammation, damage, and pain are not simply caused by wear and tear but by an attack by the body's immune system on joint tissue, similarly compared four weeks of 40, 100, or 200 mg SC-58635 twice a day to a placebo in 330 patients. Patients treated with SC-58635 experienced improvement in their assessment of their symptoms, less morning stiffness, and fewer painful and tender joints (Lipsky and Isakson 1997).

WE THINK WE MAY HAVE A DRUG. WHAT'S NEXT?

The Phase 2 trials for copolymer-1 and celecoxib, though quite different in scope, complexity, and number, were similar in that both programs provided the evidence that each company needed. Before Teva signed the license for copolymer-1 and after Searle ran and analyzed the Phase 2 studies of celecoxib, each had the evidence that they needed that they had a drug. The next steps are to plan the Phase 3 studies and meet with the FDA. An end of Phase 2 meeting with the FDA to review the results of the Phase 2 trials, the longer-term animal safety studies, and the plans for Phase 3 studies allow the development of an agreement between the FDA and the sponsor about the overall plan, providing a level of confidence that if the filing adequately documents the studies, the FDA will accept the application for review (FDA 2000a).

11

Phase 3 Testing

The next step on the path to approval for a new drug is the performance of the Phase 3 trials—larger, well-controlled, randomized, and blinded clinical trials—to provide evidence of the effectiveness and safety of the drug. Phase 3 trials not only expand the testing to larger numbers of subjects but also are intended to more closely mimic the pattern of patients seen by physicians. The Phase 3 trials always include pivotal trials that are critical for the demonstration of effectiveness and safety and may also include other smaller supportive trials designed to extend the claims for the drug; for example, to demonstrate utility in special groups of patients such as the elderly or children, or to provide evidence of enhanced safety. Both types of trials are the key determinants for the market of the drug, but the drug will never be approved without successful pivotal trials.

The FDA encourages sponsors to have an "end of Phase 2 meeting" with FDA staff to map out the studies that will be needed to file for approval. The purposes of the meeting are "to determine the safety of proceeding to Phase 3, evaluate the Phase 3 plan and protocol . . . and identify any additional information necessary to support a marketing application" (CFR 2003b). The goal of the meeting is to establish "an agreement between the FDA and the sponsor of the overall plan for Phase 3 and the objectives and the design of particular studies" (CFR

2003b). The minutes of the meeting serve as a permanent record of the agreement and "barring a significant scientific development that requires otherwise, studies conducted in accordance with the agreement shall be presumed to be sufficient in objective and design for the purpose of obtaining marketing approval for the drug" (CFR 2005j). Simply put, this meeting establishes a Phase 3 plan acceptable to the FDA's reviewers. The willingness to have such meetings, codified in federal regulations in 1987, departed from the earlier practice of maintaining strict independence of the FDA's medical reviewers from sponsor scientists, a practice that could and did allow large clinical studies to be performed that, despite the outcome, would not be acceptable in design to the FDA reviewers (Hilts 2003).

Deciding on the design and number of the Phase 3 studies is a major challenge for the sponsor. In most companies, a project team composed of company physicians, scientists, and marketing professionals (whose job is to make sure that the drug has a large and successful market) have been working on the clinical development plan since the drug began development. The members of the team are knowledgeable about both the proposed indication(s) for the drug (the specific diseases it will treat) and the rules and science of drug testing. Although each step in the process of drug development is vital, putting together the Phase 3 plan, getting it approved by senior management, and carrying out the studies are momentous because of their importance to drug approval, the costs, and what is at stake. About one in five drugs for which an IND is filed are approved, and fewer than 30 percent of the drugs that enter Phase 1 enter Phase 3 (DiMasi 2001; DiMasi et al. 2003). Attrition is higher earlier in development, with fewer than four out of one hundred compounds that begin development being approved (Shillingford and Vose 2001), but late failures are not unknown.

The overall strategy for the development of a drug is first outlined when the decision is made to begin development, when the company projects the cost of the research and the potential for success not only in gaining approval but also in selling enough of the drug to recover its costs and make a profit. The costs and risk of each step in development are high. On average, Phase 1 trials for a compound cost over $30 million and take two to three years to complete; about 70 percent of drugs will move to Phase 2. Phase 2 costs over $40 million and generally requires two to four years to complete. Fewer than half of the drugs that begin Phase 2 move to Phase 3. The mean out-of-pocket expenses for Phase 3 trials for a single new drug have been reported to be $86 million (DiMasi et al. 2003). When this figure is capitalized, corrected for time and the cost of money, it rises to $119 million (DiMasi et al. 2003).

In developing the Phase 3 plan, the project team must define, based on the results in the lab and in the Phase 1 and 2 studies, the indications, which are the precise medical conditions for which the drug will be used and thus tested for. This can be a complex problem, particularly if there are many other approved drugs for that condition. Will the sponsor's drug be useful for all the patients with the condition or only a subset that have a particular form or stage of the disease? Will the sponsor's drug be approved only for patients who have failed with X, Y, or Z drug? If the team elects to seek approval for a narrow subset of patients with a certain condition, then the market for the drug may be too small to make financial sense for the company. If they seek the widest use, for example everybody with arthritis, they are at a greater risk of failing to demonstrate safety and efficacy and therefore failing to get approval. This is based on biology. Diseases and people with diseases are varied, and taken in the largest, least discriminatory way, they may vary a great deal in their response to the drug. As discussed earlier, people respond to a stimulus, such as a drug, over a range. Some may have a large response, some small, but the response of the population will generally follow a bell-shaped curve. A group made of individuals more like each other will produce a narrower curve. Clinical trials test the idea that the curve for responses to the test drug is statistically separate from the curve of the response to the control drug. The null hypothesis is that all subjects are actually part of the same curve. If the drug is tested in patients that are more varied, then that variation increases the risk that the curve for each group is wide and cannot be separated statistically. The drug will fail.

Biostatisticians estimate the response in the control group using data from the literature on other studies with patients with the condition. Then, based on the difference between the two groups that the team's medical experts project is necessary to gain approval, the biostatisticians calculate the size of the trial needed to generate the results that would exclude the null hypothesis. This can be a very large number of human subjects. Phase 3 trials are generally larger than Phase 2 trials, and for some diseases Phase 3 trials may require tens of thousands of participants. The costs and risk are high, so the logic and analyses about the indications and design of the Phase 3 plan are very carefully made and require approval from the highest research and business leadership in the company. The discussions around these decisions start before the development process starts and are ongoing throughout the years of animal testing and early human trials, but when the decisions must be made for the pivotal Phase 3 trials, the choices must be made. Should they develop for this specific use and that one, but not the one unlikely to succeed or unlikely to generate a sufficiently large market?

Indications with small potential markets are, not surprisingly, unattractive to pharmaceutical companies for business reasons. Congress has acted to provide incentives for companies to develop drugs that will be useful to a small number of individuals. The 1962 amendments to the FDCA required that a drug must be shown to be safe and effective for the medical use printed on the label. It became apparent in the 1970s that companies were focusing their efforts on drugs that would treat large numbers of people, in part because the large market would hold the promise for large revenue but also because the early pharmaceutical industry had developed for over fifty years with a focus on drugs for infections and for heart and central nervous system disease—the diseases that were most responsible for serious consequences, including death. Scientific advances had been greatest in those areas, and the pioneer compounds provided chemical leads and clinical strategies for antibiotics and drugs for high blood pressure and nervous system disorders (Landau et al. 1999). After World War II, there was an expansion of medical research and a growing expectation of patients, the consumers, for equally effective treatments for diseases that affected smaller numbers of people. In the 1970s and early 1980s, calls arose for changes in the FDA's drug approval regulations that would create incentives for companies to develop drugs for the serious diseases that affected relatively small numbers of peo-

ple. The response to this campaign was the passage in 1982 of the
Orphan Drug Act. The act has been amended several times since its
passage, and only in 1994 did the FDA publish the regulations
(Pusinelli 1999). The current rules of the Orphan Drug Act provide a
process by which a company can request that a drug be declared an
orphan drug. The indication for a drug designated an orphan drug
must affect less than 200,000 people in the United States or, if it affects more than 200,000, there must be *no* reasonable expectation
that the cost of developing and making the drug available in the
United States will be covered from sales of the drug in the United
States (CFR 2005o). The act provides tax incentives, extended periods of exclusivity, and grants to carry out the research needed to gain
approval (CFR 2005o). Since its enactment, over 1,455 drugs have
been designated orphan drugs, and over 231 have been approved for
sale (FDA 2005). Application for orphan drug designation is usually
made when Phase 2 trials have shown that the candidate drug has potential utility and the indication(s) can be defined with some precision. Copaxone was designated an orphan drug in 1987, soon after it
was licensed by Teva (FDA 2003b).

The challenge for the project team is to design the most efficient development plan, within the regulatory and ethical constraints, that will
provide the largest market, and the best return on investment. The trials should be no larger, nor run longer, than required to provide evidence for efficacy and safety. A poorly designed trial without the
statistical power to convince the reviewers is not only a waste of
money but also unethical because human beings would be exposed to
the risks of the trial with no benefit to them or to society. The trials
proposed to the FDA at the end of the Phase 2 meetings must convince
the reviewers that they are a robust test of the null hypothesis, the hypothesis that the control and test drugs are not different, or they will
not agree to the plan and will suggest modifications. The trial results
may or may not go the way the company wishes, showing the effectiveness of their drug, but without the documented performance of an
acceptably robust test of the null hypothesis, the application for approval may not even be accepted as filed.

The 1962 amendments to the Food, Drug, and Cosmetic Act, also
known as the Kefauver-Harris Drug Amendments, required that

sponsors show that a new drug is safe and effective for the indication on its label. But the medicine and the science of clinical trials have changed since 1962. Improved understanding of disease processes has resulted in a more precise and narrower focus on specific forms or stages of diseases. The indications for new drugs may be narrower but the number of indications more numerous. In simpler terms, as we better understand our bodies and how they might malfunction, there is a tendency to "slice and dice" diseases. And as statisticians and clinical trial specialists have gained experience, the demand for more rigorous performance and analysis of trials has grown.

The FDA provides guidelines on the topic of demonstrating effectiveness (CDER 1998a). The FDA's legal position, supported by case law, is that Congress' intent in the 1962 amendments was to require at least two adequate and well-controlled trials—two pivotal trials. Each should be performed at multiple research sites to assure that the results are not caused by something special about one hospital or research center. But two pivotal trials are not always required. The FDA is flexible and will consider information from other adequate and well-controlled studies (for example, trials with other doses, administration schedules, or endpoints) to support a single adequate and well-controlled study. In addition, a single well-designed multicenter study has been sufficient for approval when the data were found to be reliable and statistically strong and provided "evidence of an important clinical benefit, such as an effect on survival, and a confirmatory study would have been difficult to conduct on ethical grounds" (CDER 1998a). The key is to provide scientifically sound and well-documented evidence within the existing medical and ethical constraints.

The question central to the design of Phase 3 studies is what will be printed on the label about the medical condition for which the drug is approved. From the outset, the scope of the FDA's regulation has been directed at the claims on the label. Fundamentally, the laws empowering the FDA are concerned with the legality of shipping misbranded (mislabeled) or adulterated materials across state lines. Initially, it was about the contents, then safety concerns were added, and finally efficacy was captured in the 1962 amendments. On the label, the approved indication for use of the drug must be provided to physicians so they will know how best to use the drug. The sponsor must design and carry out trials that provide statistically sound evidence that patients, whose medical condition is precisely defined, benefited from a course of treatment with the drug.

The term label in this context does not mean the piece of paper stuck on the bottle or jar but the words used on the material provided to the prescribing physician that describe what types of patients have been shown to benefit from the drug. The marketing and advertising of a drug are strictly bound by the label and, as will be discussed in Chapter 16, an ongoing tension exists between drug manufacturers and the FDA about marketing and advertisement activities that stray from the label.

PHASE 3 TRIAL OF COPOLYMER-1, NOW GLATIRAMER ACETATE

Glatiramer acetate and celecoxib present two very different development situations in this final stage of drug testing for approval. Glatiramer acetate had been tested in a very well-designed trial and shown to be active. The market for a drug for MS is limited; in the United States it is estimated that somewhere between 300,000 and 400,000 people have MS; worldwide the number is thought to be 2.5 million, but the market for an expensive drug requiring refrigeration and daily injection likely will be limited to the developed world, providing a smaller market. Based on the course of the disease, the conception of how glatiramer acetate might work, and the unlikelihood of altering the established disability, the trials were designed with the endpoint of reducing the rate of relapses. This means that the indication would be limited to only those who have the relapsing forms of MS. About 85 percent of patients with MS have the relapsing pattern at diagnosis, but the disease progresses in a significant fraction of the patients to a form without relapses. For Teva, from a financial perspective, this was both bad and good. It was bad because of the resulting small market of less than 200,000 in the United States and good because it allowed them to apply for and receive orphan drug status in 1987. Aiding the plan was the fact that Teva had a clear indication from the

FDA that a second, larger trial with a design similar to that of the Bornstein trial would be sufficient.

Thus, only two trials of glatiramer acetate were pivotal for U.S. approval, the pilot trial of Bornsten et al., described in Chapter 10, plus a larger, long-term placebo-controlled trial in relapsing-remitting MS (CDER 1996a; Johnson et al. 1995). In this second trial, supported by grants from the FDA's Orphan Drug Program, the National Multiple Sclerosis Society, and Teva, 251 patients with documented relapsing-remitting MS in multiple U.S. sites were randomized to daily self-injections of either glatiramer acetate (125) or a placebo (126) for twenty-four months. The primary endpoint was the proportion of relapse-free patients during the twenty-four–month trial. Secondary endpoints were the frequency of relapses, change in the disability score from baseline, proportion of progression (change in disability score by at least one unit, sustained for three months or more), and time to progression. Both the design and endpoints had been agreed upon during meetings between the Teva Pharmaceutical representatives and FDA staff reviewers (CDER 1996a; Pinchasi 2003). The results of this study supported the conclusion of the Bornstein et al. study. Thirty-three percent of the patients treated with glatiramer acetate were relapse-free, compared with 27 percent of the placebo group, a difference that does not reach statistical significance, but the number of relapses in the control group (1.68/patient for two years) was significantly higher than that of the copolymer-treated group (1.19/patient for two years). Also significant was that overall there was a tendency for a slight improvement in disability measures in the patients treated with glatiramer acetate and a worsening in the control patients (Teva Neurosciences 2002). A subsequent international study, required by the European regulators, compared the number of active brain lesions, visible by monthly magnetic resonance imaging, in 119 patients receiving glatiramer acetate with that of 120 patients on a placebo over nine months. This endpoint has gained credence as a measure of MS disease activity (Harris et al. 1991). The difference in the number of active lesions seen during the nine months between the copolymer group and the placebo group was highly statistically significant, favoring glatiramer acetate (Comi et al. 2001; Teva Neurosciences 2002).

In all studies, the only adverse events were frequent mild to moderate injection site reactions of redness, inflammation, and pain plus the systemic reaction, consisting of short-term flushing, chest tightness, palpitations, and feelings of anxiety at least once in about 10 percent of patients receiving glatiramer acetate.

PHASE 3 TRIALS OF CELECOXIB

The Phase 3 plan for celecoxib was more extensive because several indications were being tested and because the number of people who would be taking the drug, were it approved, would be large and they would be much less seriously ill. Five pivotal randomized, double-blind trials of celecoxib in osteoarthritis were conducted. Two pivotal twelve-week trials for osteoarthritis of the knee and one for osteoarthritis of the hip compared celecoxib taken twice a day at 50, 100, or 200 mg with placebo and naproxen sodium, a conventional, nonselective COX inhibitor, at 500 mg taken twice a day. In one of the twelve-week trials, the incidence of ulcers in the stomach and the beginning of the small intestine was determined by performing endoscopy at the beginning of the trial and at 12 weeks. Also, two pivotal six-week trials of OA of the knee compared placebo with celecoxib at 100 mg taken twice a day or 200 mg of celecoxib taken once a day. A total of 4,661 subjects were enrolled in these five trials, 2,880 of whom took celecoxib. The primary endpoints for the trials were four different assessments, three by patients and one by the study physicians, that provide information on how the patient and physician perceive the impact of arthritis on the patient's functioning and the amount of arthritis pain and stiffness the patient is experiencing. Each of these assessments or indices had been developed and tested by different groups of physicians, and the use of these four was agreed to by the sponsor and the FDA to provide the best confirmation for the subjective endpoints of pain and stiffness (CDER 1998). In both twelve-week studies, celecoxib at all doses was better than the placebo and comparable to naproxen. In the six-week study comparing one higher-dose pill taken once a day with two lower-dose pills taken in the morning and evening, celecoxib was superior to a placebo and there were no differences in the assessment scores between the two schedules for taking celecoxib.

Two pivotal randomized, double-blind trials in rheumatoid arthritis were done. Both were twelve-week, five-arm trials comparing celecoxib at 100, 200, or 400 mg twice a day with placebo or 500 mg of naproxen taken twice a day. A total of 2,257 patients were entered into these trials and were nearly equally distributed into the five arms. As in the OA trials, the primary and secondary endpoints were based on a series of assessments completed by the patient or physician. A rheumatoid-arthritis–specific assessment was also used that included laboratory tests for proteins found in the blood during active joint inflammation and X-ray measures of joint swelling and inflammation. Considering the patient and physician global assessments and the count of tender or painful joints, celecoxib at all doses was superior to a placebo and comparable to naproxen. More people withdrew from the study for lack of efficacy in the placebo group than in either the celecoxib or naproxen groups. The lab tests for blood signs of joint inflammation did not give clear-cut results, in part because the subjects varied widely at the beginning of the study.

In all pivotal and supportive studies, celecoxib was reported to be well-tolerated. The detailed analysis of safety will be presented later when the entire submission is discussed. In the pivotal arthritis trials in which endoscopy was done to detect stomach ulcers or damage to the lining of the stomach or small intestine, such findings occurred less frequently in patients receiving celecoxib than in those receiving the nonselective COX inhibitor naproxen (CDER 1998b).

Overall, these studies indicate that celecoxib is superior in efficacy to a placebo for the treatment of osteoarthritis and rheumatoid arthritis and, in both settings, comparable with naproxen, a previously approved COX inhibitor.

12

Putting Together the Application for Approval: The New Drug Application (NDA)

The New Drug Application (NDA) is the document on which the FDA bases its decision that the drug is safe and effective, that the drug's benefits outweigh its risks, that the labeling is appropriate, and that the drug is manufactured so that its identity, purity, strength, and quality are preserved (Mathieu 1994). The formidable task of the team assembling the NDA is to provide the information, the data, and the analyses of the data to support approval. Although assembling the NDA begins in earnest when the clinical trials in Phase 2 or Phase 3 indicate that the drug is likely to meet the efficacy and safety objectives, the planning work begins years earlier. All along, the project team has been tracking the animal safety studies, the progress of the clinical trials, and the manufacturing of the drug. By the time the Phase 3 trials are under way, the laboratory and animal safety study reports have been written and checked out by the quality assurance department for accuracy and completeness. The reports of studies documenting the purity, consistency, and stability of the drug substance and the formulated drug product have also been written and checked out. Also, the reports of the early human studies to find out how the human body takes up, breaks down, and excretes the drug will have been written and reviewed by the quality assurance department. But only when the later human trial data suggest that the drug is a "go" does it make sense to expend the time and effort to assemble all the required efficacy and human safety study reports and documentation and do the data analyses and writing for the final document.

Assembling the NDA takes substantial time and human effort, perhaps a year or more of work by dozens of individuals. Most companies assign the responsibility for shepherding the process to a committee similar to or identical to the project team. Members include experts in the manufacturing and quality control of the drug substance and the formulation and manufacture of the drug product, pharmacologists, toxicologists, clinicians, biostatisticians, and staff from the regulatory affairs department with knowledge and experience in the requirements and working of the FDA, as well as legal, marketing, and business staff. Companies manage the process in different ways, but one model has a regulatory submission project manager and a clinical submission project manager working together to plot out the timing of each task and to call and run the meetings of the team to identify and solve problems and possible delays as the work progresses (Francher, personal communication 2004). Whether there is one or a pair of team leaders, the task requires skill and experience in management of complex projects, diplomacy, knowledge, and the ability to run meetings. And there are lots of meetings. They may be once or twice a month in the beginning and then more frequent, with two or three each week as the filing date approaches. Once the final work begins, after Phase 2 or 3, a projected filing date is set and senior management is kept up to date on the progress, especially anything that suggests problems or delays.

During the clinical trials, detailed information about each test subject is collected: the medical information documenting that the patient met the eligibility requirements for the study as described in the protocol, the response to the drug, and the results of physical or other exams and laboratory tests as required by the protocol. The staff of the medical center doing the study enters the information in a document standardized for each study, called a case report form (CRF). In the CRF, the patient is not identified by name but by his or her hospital number. The completed CRFs are submitted to the company, either directly into a database for the study or by company data-entry clerks. While each study is going on and at its completion, company study monitors will visit the medical center and document that the information in the case report form matches the patient's medical or hospital chart. The hospital and physicians' office charts are the legal documents of record for treatment of a patient, study participant or not, and the data used for the analyses and submitted to the FDA must match the chart. Quality assurance staff will also perform audits to check that the data submitted to the company match the information in the patient's chart. When the pivotal trials are complete, including the follow-up as required by the protocol, the database of case report forms of all the humans who have taken the drug is analyzed for completeness and validity (consistency on audit with the hospital charts). When the database is approved by the reviewing physicians and by quality assurance, it is locked down, made unchangeable. This allows the physicians analyzing the results of all pivotal and supporting trials for efficacy and the physicians assessing the safety data on all humans treated to begin their analyses.

Each pivotal and supporting study is analyzed for efficacy, and an integrated efficacy summary is prepared using all these trials. The integrated safety summary provides the information relevant for assessing the safety of the drug based on *all* human beings exposed to the drug, not just the subjects in the pivotal trials.

Generally, the entire clinical database is not submitted to the FDA, just the data on all of the subjects in the pivotal trials and the file of every subject who experienced a serious adverse event, including all deaths, whether or not the adverse events are thought to be caused by the drug. Once the safety database is cleaned and locked, the information on the adverse events is organized into tables based on which part of the body was affected and how, using standard MedDRa coding language agreed upon by the FDA and regulatory authorities in Europe and Japan (ICH 2003a).

The FDA provides instructions for completing the NDA that detail not only what must be included but also how it should be organized and transmitted to the FDA. As a paper document, an NDA can run to several hundred thousand pages arranged in hundreds of volumes. In the past, conventional NDAs required trucks for delivery to the FDA. In the 1990s, the FDA began to develop procedures and guidelines to allow electronic filing of NDAs and have provided a guidance document for electronic submission (CDER/CBER 1999). Electronic submissions have several advantages, not the least of which is the ability to use hyperlinks for cross-referencing data entries and reports. This allows, for example, a reviewer to go directly to a particular study report or the case report form for a particular patient during the review of the integrated safety summary. Initially the FDA did not specify the software to be used for reports and data files with electronic submissions, a situation that was not ideal. Beginning with the publication in 1997 of regulations that "provided for the voluntary submission of parts or all of regulatory records in electronic format without an accompanying paper copy," the FDA has worked to provide guidance for electronic submissions to speed the process of submission and review (CDER/CBER 1999). One hurdle is the part of the FDCA that requires signatures on documents submitted to the FDA to meet regulatory requirements. This applies to many forms, from the forms filled out by clinical investigators to the cover sheets on the NDA. The current regulations provide rules on the procedures needed to establish and protect the individuality and validity of electronic signatures so that electronic signatures have the same legal, contractual weight as handwritten signatures (CFR 2003d).

The current e-submission guidance is intended to provide industry with the instructions for standard software (PDF, XML, and SAS) and media (floppy discs, CDs, and tape) that will allow the reviewers access to the needed reports and data (CDER/CBER 1999).

THE COPAXONE (GLATIRAMER ACETATE) NDA

The NDAs for the drugs we are tracking bridged the transition to complete electronic filings. The NDA for copaxone was filed on June 15, 1995, little more than a year after the last observation of a patient in the second pivotal trial. The efficacy data were limited to the results of the two placebo-controlled trials, enrolling 51 and 251 patients, respectively, plus an extension of the second trial, which followed 125 patients (CDER 1996a; Johnson et al. 1995; Johnson et al. 1998). Scientifically, each trial is analyzed independently for efficacy and, as described earlier, the Bornstein et al. study showed a highly significant effect on the frequency of relapses and the proportion of relapse-free patients. The second, larger, multicenter study confirmed these results, but the effect was statistically small.

The integrated safety summary (ISS) of glatiramer acetate provided safety information on a total of 1,008 individuals with MS involved in over eleven clinical trials, 901 of whom received one or more doses of glatiramer acetate (CDER 1996a). Seven hundred and seventy-nine people with the relapsing form of MS had received the drug, 670 for at least six months, 490 for at least twelve months, 290 for at least two years, 87 for at least three years, and 15 for at least five years. There were twice as many females as males (remember that MS occurs much more frequently in women than in men) and, though the age ranged from 18 to 68, the average age was 30, consistent with the demographics of MS patients. There were seven deaths, all in patients receiving glatiramer acetate, and though none were attributed to glatiramer acetate, in two there was an association with what was called the systemic reaction: the flushing, heart palpitations, and chest pain that resolve in a few minutes. Eighty-seven of 844 subjects in the controlled clinical trials reported experiencing the systemic reaction at least once and most (52 of 87) only once. Serious adverse events occurred in 6.5 percent of subjects receiving glatiramer acetate and 6.8 percent in the placebo group. Seventeen patients were reported as having had a serious adverse event possibly related to the study drug. These affected the body as a whole (for example, a rash or fever and chills) or the circulatory, digestive, nervous, and blood systems, but there was no pattern; in fact, no one adverse event was seen in more than one patient. As might be expected for a group of this many individuals, more than fifty patients experienced serious events considered unlikely to be related to the study drug.

When all trials are considered, 8.1 percent of the subjects receiving glatiramer acetate decided to drop out of a trial, but in the two pivotal trials, the percentage of subjects dropping out was the same in both arms, ~21 percent. Adverse experience, specifically injection site reactions of pain, inflammation, itching, and swelling, was the single most frequent reason for the glatiramer acetate subjects to drop out.

Only four of eighteen dropouts among the placebo group resulted from adverse events; most were simply attributed to the subject's decision, with no details given.

Laboratory tests of blood, urine analyses, electrocardiograms, and vital signs did not show clinically significant changes. The safety and efficacy significance of the development of antibodies to glatiramer acetate is not known, but most subjects receiving glatiramer acetate did develop antibodies to it. The antibody level dropped to background after three to six months. Beyond the infrequent and self-limiting systemic reaction and injection site reactions, no significant adverse events attributed to the drug were detected in the safety analysis of glatiramer acetate.

THE CELEBREX (CELECOXIB) NDA

The NDA for celecoxib was filed electronically on June 29, 1998. A paper copy was also filed and amounted to 450 volumes. The clinical efficacy and safety portions of the celecoxib NDA reported the results of fifty-one trials, enrolling over 13,000 subjects (CDER 1998c). Searle was seeking approval for the drug for three indications: osteoarthritis (OA), rheumatoid arthritis (RA), and management of pain.

There were five pivotal studies in osteoarthritis: two twelve-week studies in OA of the knee, one twelve-week study in OA of the hip, and two six-week studies in OA of the knee. The twelve-week OA studies comparing celecoxib at 50, 100, 200, and 500 mg twice a day with naproxen and a placebo enrolled over 3,500 patients. All the twelve-week studies showed that celecoxib at doses at or above 100 mg twice a day was superior to a placebo and comparable to naproxen. The three six-week OA studies in 1,404 subjects compared 100 mg of celecoxib twice a day with 200 mg once a day and with placebo in patients with a flare of OA of the knee. The patient's assessment of pain and the measure of physical function were significantly improved in both celecoxib arms when compared with the placebo arm. Searle proposed that the studies showed that celecoxib was effective in OA, with efficacy similar to naproxen. The recommended dose was 200 mg per day, taken in a single dose or in divided doses.

Two 12-week studies in a total of 2,252 subjects with RA compared celecoxib at 100, 200, or 400 mg twice a day with a placebo and with 500 mg of naproxen. Celecoxib at all doses was superior to a placebo and similar to naproxen. The recommended dose was 100 mg twice a day, though they suggest that some patients may benefit by increasing the dose to 200 mg twice a day. The six-month study showing that celecoxib was superior to diclofenac in reducing the number of tender and painful joints, though no different from diclofenac in reducing swollen joints, supported the idea that the effect of celecoxib was sustained.

The studies submitted to support the pain indication included three single-dose studies on acute pain after dental surgery, the one supporting study with multiple doses after orthopedic surgery plus the three pivotal studies in acute flare of OA described previously. The three single-dose dental pain studies supported that celecoxib was superior to a placebo, similar to ibuprofen, but inferior to naproxen.

The study of the efficacy in postorthopedic surgery pain indicated that celecoxib was superior to a placebo and comparable to darvocet, a widely prescribed pain reliever with acetaminophen. Looking at the relief of pain of an OA flare, celecoxib at doses above 100 mg twice a day was superior to a placebo and comparable to naproxen. The recommended dose for pain was 100 or 200 mg twice a day.

The safety database for the celecoxib submission contained 18,439 records, representing data on 13,072 individuals. There were 12,845 records for people who had received celecoxib, 2,450 who had received a placebo, and 3,343 who received an active control, such as naproxen or ibuprofen. Serious adverse events, including effects on the gastrointestinal and cardiovascular systems, were no more frequent in the celecoxib and active control groups than in the placebo group. No deaths occurred among the placebo group, and the death rate was 0.5 per 100 patient years in the celecoxib group and 0.7 per 100 patient years in the active control group. The death rate is expressed this way to account for the length of time the subject was exposed to a drug. In the North American arthritis trials, accounting for 8,723 subjects, 4,761 of whom received celecoxib, subjects who received doses of celecoxib below 400 mg twice a day were no more likely to have experienced adverse events taken as a whole than those receiving the placebo. Those who received an NSAID were slightly more likely to have experienced an adverse event, and this difference from the placebo group was statistically significant. Stomach upset and diarrhea were slightly elevated in the lower doses of celecoxib. Stomach upset, abdominal pain, and nausea were significantly elevated in the subjects receiving an NSAID when compared with either a placebo or celecoxib at any dose, resulting in a larger number of subjects withdrawing from a study because of these NSAID effects. Headache was the single most frequent (16 percent) adverse effect of celecoxib in both the North American arthritis trials and in the group who continued to receive celecoxib in the long-term open label trial. Other adverse events seen in more than 5 percent of subjects receiving celecoxib were upper respiratory tract infections, upset stomach, sinusitis, diarrhea, accidental injury, abdominal pain, and nausea. Laboratory tests for evidence of liver or kidney damage showed results similar to the placebo and less change in liver enzymes than with diclofenac.

General safety issues are always a major focus of drug trials, but for celecoxib the ability to show that this COX-2 inhibitor was associated with less toxicity than the NSAID nonspecific COX inhibitors might provide for a better safety label and therefore an improved marketing position. Differentiation in the label, whether for efficacy or safety, is the game in seeking approval, and the data did not suggest that celecoxib was more effective than the standard NSAIDs prescribed for OA or RA. Toxic effects of the NSAIDs, particularly damage to the lining of the gastrointestinal tract and increased risk of GI bleeding, are a serious problem for those taking NSAIDs for any extended period of time (Insel 1996). These adverse effects are thought to be caused by the presence of the major form of the COX enzyme, COX-1, in the gastrointestinal tract, platelets (the small blood cell fragments involved in blood clot formation), the kidney, and many other tissues. Initial studies had indicated that the COX-2 was present only in immune system cells actively involved in inflammation, suggesting that adverse effects of the NSAIDs would not be seen. Further research had shown that COX-2 was present not only in inflammatory cells but also in the kidney, the brain, and in cells of the female reproductive system, so achieving differentiation would require solid evidence of an improved safety profile for celecoxib in adverse effects involving the platelets, the kidney, the brain, the gastrointestinal tract, the stomach, and the intestines.

A great deal of evidence for the relative safety of celecoxib specifically for these organ systems was provided in the NDA. Considered together, adverse events in the brain or psychiatric adverse events were significantly less frequent in subjects receiving celecoxib or NSAIDs than in those receiving a placebo. Headaches were also less frequent in the celecoxib and NSAID groups. A small but statistically significant increase (from 0.8 percent to 1.1 percent) was seen in the incidence of muscle spasms in the celecoxib group, though leg cramps were more frequent with NSAIDs (0.9 percent versus 0.5 percent in the placebo group). Three approaches were used to test for effects on the kidney: analysis of the frequency of kidney adverse events; analysis of effects on blood pressure, an early sign of kidney malfunction; and a thorough seven-day study of the effects on kidney function of celecoxib versus a placebo and naproxen in healthy elderly individuals and in people with chronic kidney disease. The only physical evidence of a possible adverse effect on the kidney in the twelve-week arthritis trials was a small increase in the incidence of swelling of the limbs caused by fluid retention in the celecoxib and NSAID groups. Blood pressure was not affected by either active treatment. In the pharmacology study of kidney function, no effect was seen on the glomerular filtration rate (GFR), a standard measure of the ability of the kidney to filter the blood and produce urine. A reduction in the level of sodium in the urine was seen in the first two days of the seven-day dosing with both celecoxib and naproxen.

Lack of effects on platelets was shown by analysis of bleeding-related adverse effects in the arthritis clinical efficacy trials and in the results of five pharmacology studies in healthy adults given various doses of celecoxib, NSAIDs, or a placebo for one to ten days with blood studies to measure platelet function and the clotting ability of the blood. In the arthritis trials, the incidence of bleeding-related adverse events, such as anemia or bruise-like areas of skin, were similar in those receiving a placebo or celecoxib and were significantly elevated in patients receiving NSAIDs. In the platelet pharmacology studies, laboratory tests showed no effects on platelet function even at twice the proposed therapeutic dose of celecoxib.

The relative safety of celecoxib for the gastrointestinal tract was a major focus of the safety data provided in the NDA in an effort to provide differentiation from NSAIDs. Chronic use of NSAIDs increases the risk of serious gastrointestinal events threefold, probably because the prostaglandins synthesized by the COX enzymes protect the lining of the stomach from damage by stomach acid (Insel 1996). The data submitted to support the claim that celecoxib is less likely to cause GI symptoms and damage were drawn from the adverse events reported in the arthritis clinical efficacy trials plus the results of endoscopy in five arthritis trials and the thorough analysis of complications resulting from stomach ulcers in the safety database. Subjects receiving celecoxib in the North American arthritis trials were less likely to experience indigestion, abdominal pain, nausea, or diarrhea than those receiving an NSAID. Indigestion and diarrhea were slightly though statistically significantly more frequent in celecoxib-treated subjects than in those receiving a placebo. Gastrointestinal adverse events leading to withdrawal from a study were not elevated in the celecoxib group but were increased in those receiving an NSAID. Because NSAIDs are known to reduce the level of prostaglandins in the stomach lining and increase the risk of stomach ulcers, five arthritis studies included the examination of the stomach lining for ulcers by endoscopy. Subjects in two twelve-week studies comparing celecoxib at several doses with a placebo or naproxen had endoscopy before treatment began and at week twelve; both studies found a significant increase in stomach ulcers with NSAID treatment but not with celecoxib. Also, more ulcers were seen with the NSAID diclofenac than with celecoxib after six months in the six-month RA trial. To track stomach ulcers that may appear and heal, subjects in two twelve-week arthritis studies had an endoscopy performed every four weeks. One study was done in OA patients and compared celecoxib (200 mg twice a day) with naproxen; the other was done in RA subjects and compared celecoxib (200 mg twice a day) with diclofenac and ibuprofen. The results in the 1,636 subjects showed that ulcers were more frequently found in patients receiving ibuprofen than in those receiving either celecoxib or diclofenac. Ulcer complications, including bleeding, perforation, or obstruction of the passage between the stomach and small intestine, were tallied in 11,008 patients in fifteen studies. In fourteen controlled trials in which celecoxib was being compared for safety and efficacy with a placebo and an NSAID, the annual incidence was not significantly different between a placebo and celecoxib and was significantly increased in those subjects receiving an NSAID. The incidence of these ulcer complications was essentially the same for patients receiving celecoxib in the controlled trials and the open label extension of the OA trial, supporting the contention that the low risk versus NSAIDs was sustained over longer-term celecoxib treatment. Based on these results, Searle concluded and proposed to the FDA that the celecoxib label need not carry the NSAID warning of the risk of GI toxicity.

NOW THE REGULATORS DECIDE

With the filing of the NDA, the future of these drugs depends on decisions made by the reviewers and regulators at the FDA. Will the NDAs pass muster and do so with a label that allows the sponsors to sell enough drug to recover the development costs and earn a profit, so that stockholders are rewarded with a dividend and funds are available to support the discovery and testing of new drugs? The answers to these questions consume the efforts of many; not just FDA reviewers but also company scientists and regulatory staff. The process of reaching the answers may take months or years, and there are no certainties, except anxiety and stress for the scientists whose years of work now face questions, tests, and judgment.

13

Everybody Holds Their Breath: Will the FDA File the NDA?

The next hurdle for a company seeking approval of a new drug is the FDA's acceptance of the NDA filing, a step achieved smoothly for some but with much public drama by others. Witness the Erbitux filing by ImClone in October 2001 and the financial scandal that played out not only in the pages of the *Wall Street Journal* but also in nearly every media outlet in the United States after the FDA refused to accept the filing for Erbitux on December 28, 2001. In the ensuing months, the CEO of ImClone and a lifestyle advice guru and media magnate had more than their fifteen minutes of infamy. All of this was set off by the FDA's refusal to accept the Erbitux filing, which was actually a biological license application (BLA), required at that time for an application to sell a biologic medicine such as Erbitux. The financial skullduggery of some in ImClone's management that brought the shrill media attention deserves more probing of character weakness than fits in this venue, but exploring the reasons given by the FDA for the refusal to file the Erbitux BLA will illuminate several issues central to the complex process we are examining.

About six to nine months prior to an NDA submission, the FDA, based on the "end of Phase 2" meeting, the agreements resulting from that meeting, and annual progress reports received from the sponsors, will schedule a pre-NDA meeting. For this meeting, the sponsor is asked to prepare a briefing document (Kukich 2003). Based on that document, an FDA review team for the anticipated NDA will be set up. The review team will generally include (FDA 2002a):

- chemists and microbiologists to review how the drug is manufactured;
- pharmacologists and toxicologists to evaluate the effects of the drug on laboratory animals;
- physicians to review the results of the clinical (human) trials;
- clinical pharmacologists and biopharmacologists to review the interaction between the body's response and the drug dose and how the body takes up, distributes, metabolizes, and eliminates the drug;
- statisticians to evaluate the design and results of the clinical studies; and
- project managers to "orchestrate and coordinate" the drug review team's interactions and reviews and who serve as experts on the regulations and are the primary contact between the sponsor and the FDA.

When the NDA is submitted, the review team goes over the submission to determine whether it is suitable for review and may be filed, a decision that must be communicated within sixty days of receipt of the NDA (CDER 1993).

The formal reason for which the FDA may refuse to file (RTF) is, as briefly stated in the law, failure to file a completed application form. An application is incomplete when it does not on its face contain information required by law (CDER 1993). In their 1993 guidance, CDER described in detail the three circumstances in which it has exercised its RTF authority:

1. Omission of a section required "or presentation of a section in so haphazard a manner as to render it incomplete on its face" (CDER 1993). The guidance lists a comprehensive table of contents, a summary, the technical sections, and required case report forms and tabulations.

2. Clear failure to include evidence of effectiveness; for example, lack of adequate and well-controlled studies, "presentation of what appears to be only a single adequate and well controlled trial without adequate explanation of why the trial should be regarded as fulfilling the legal requirements" (CDER 1993), use of an inappropriate study design, and for combination drug products "failure to present studies that assess the contribution of each component" (CDER 1993).
3. "Omission of critical data, information or analyses needed to evaluate effectiveness and safety or provide adequate directions for use" (CDER 1993). The guidance provides examples of omissions that have led the FDA to exercise its RTF authority, which is essentially a list of the required contents of an NDA.

The message of the guidance is clear; the law, supplemented by the various FDA guidance documents, lays out what is required for a reviewable NDA. If on the initial inspection FDA finds that the application is so deficient that it is not potentially approvable, the FDA will exercise its authority to refuse to file because it wastes the agency's resources and diverts reviewers from reviewing complete applications.

The contents of the NDA should not surprise the FDA. By statute and practice, the FDA has regular meetings with developers of new drugs, including a pre-IND meeting, end of Phase 2 meetings, and a pre-NDA conference. As mentioned in Chapter 11, the minutes of such meetings, particularly the end of Phase 2 meetings, essentially are an agreement between the sponsor and the FDA (i.e., if the sponsor carries out the agreed studies and they provide evidence of efficacy and safety, the FDA will find the studies sufficient). It is simple: do what we all agreed is necessary and we will accept that as sufficient. This does not guarantee approval but just assures review if the document is complete and the data consistent from section to section.

WHAT WAS BEHIND THE ERBITUX STORY?

Why did the FDA refuse to file the Erbitux application? Details of RTF decisions may or may not become public knowledge, but in this case they did, in part because ImClone had worked very hard to raise expectations for what was seen as a first-in-class monoclonal antibody to the epidermal growth factor receptor (EGFR) to be developed. Also, in September 2001, Bristol-Myers Squibb agreed to pay $2 billion for 20 percent of ImClone stock and the marketing rights for Erbitux in the United States, Canada, and Japan (Johnson 2001). ImClone would receive $1 billion on reaching certain development milestones and keep 60 percent of sales of Erbitux after approval. This purchase at $70 per share was more than a 35 percent premium over the stock price at that time, suggesting that Bristol, known for the successful development and marketing of many cancer drugs, had high expectations for Erbitux. The stakes were high financially, and the physicians and cancer patients were expecting a safe and effective drug for patients for whom all other remedies had failed.

The RTF decision was communicated to ImClone in a letter dated December 28, 2001, and after the furor was raised and ImClone did their spin, the letter was published in its entirety in *The Cancer Letter*, a Washington-based newsletter for oncology researchers, on March 8, 2002 (Anonymous 2002). The description of the reasons for the RTF decision is long and detailed, focusing on six issues surrounding the clinical trial data, particularly the pivotal trial in patients with advanced colon cancer. The letter first states that the application is scientifically incomplete because it does not contain data that isolates the contribution of the chemotherapy agent given with Erbitux, does not provide data to support the proposed dose, and does not provide data to show that the subjects receiving the combination were unresponsive to the chemotherapy agent alone. The agent being used with Erbitux is irinotecan, a chemotherapy drug with some activity in colon cancer but with significant toxicity. The design that the FDA had agreed to for the pivotal trial included a group of patients receiving Erbitux alone. Excluding that arm prevented a determination of whether the patients treated with the combination for which ImClone was seeking approval were being needlessly exposed to the toxic agent irinotecan. The letter also refers to a meeting in August 2000, where the FDA stated that documentation of irinotecan unresponsiveness in patients entering the trial is "crucial" and notes that the application failed to provide documentation by the independent committee established to review the patients' CT scans that confirm unresponsiveness to irinotecan.

The second reason given for the RTF was the substantial deviations from the pivotal trial protocol documented by ten specific patients who did not meet the protocol eligibility requirements or whose treatment during the study deviated from the plan in the protocol. Ten deviations in a trial that enrolled 120 patients can have a large impact on the ability to draw conclusions from the trial.

The third reason for the RTF was the discrepancies and deficiencies in the data sets for the primary efficacy analysis, including differences of opinion among radiologists reviewing the CT scans for irinotecan unresponsiveness and a failure to indicate the computer program used to determine whether a patient had responded, particularly where the response score provided was not consistent with the radiologist's score.

The letter also notes that the safety database is incomplete and inconsistent so that there cannot be an "accurate assessment of the toxicity profile." Also, patients were entered into the efficacy trial one month before the protocol was submitted to the IND. This is a violation of the law that prohibits treatments with an unapproved drug outside of an approved clinical trial under an open IND.

Finally, the letter states that, given the deficiencies, the study did not provide an adequate sample size or a "clinically meaningful overall response rate," thus precluding approval based on a single study. It is highly likely that the number of appropriate subjects and response rate needed for approval based on a single study had been agreed to by the FDA and ImClone and that the deficiencies meant that ImClone had missed the agreed upon mark.

The letter makes it clear that FDA staff had carefully reviewed the filing and documented that it did not contain the data that might support approval. This does not appear to be a capricious decision but rather one based on a detailed comparison between what was "promised" by ImClone and the discrepancy-laden filing.

One curious point is why the agreed upon trial design, Erbitux alone versus Erbitux plus irinotecan, was not adhered to? It may relate to a patent issue. ImClone does not hold patent protection for the use of an antibody to EGFR alone to treat cancer. The large biopharmaceutical company Genentech holds a patent that claims the use of an EGFR antibody alone to treat cancer. The patent held by ImClone covered the use of an EGFR antibody plus a chemotherapeutic agent (Schlessinger et al. 2001). This may have caused ImClone not to include an Erbitux-alone group, perhaps fearing that Erbitux alone would be found as effective as the combination and without as much toxicity. This situation would require that they pay to license the Genentech patent if they wanted to sell Erbitux, an expensive consequence of success (Dove 2002).

In any case, the BLA was not accepted for filing, and financial, legal, and political drama ensued. After extensive discussions with the FDA, Bristol-Myers Squibb and ImClone submitted in August 2003 a revised BLA that included data from trials performed by their German partner, Merck K.G.a.A., which is unrelated to Merck & Company in the United States (Bloomberg 2003). This BLA was accepted for filing, and because Erbitux received fast-track status, the FDA's decision was due in February 2004.

On February 12, 2004, the FDA announced that Erbitux had been approved for sale for the treatment of EGFR-expressing metastatic colorectal carcinoma in patients who are refractory to irinotecan-based chemotherapy. Although the approval was good news, the delay of twenty months provided a window for the completion of clinical development of another drug, Avastin (by Genentech), to be used along with chemotherapy drugs for patients with colorectal cancer that had spread to other organs. Unlike Erbitux, Avastin received approval as a "first line" therapy with any chemotherapy combination that included 5 fluorouracil, an advantageous label that supports use in more patients and with more flexibility in the choice of chemotherapy drugs (Pollack 2004). It will be months or years before the positions of Erbitux and Avastin in the care of patients with colorectal cancer are known. But ImClone's gamble may well have lost them the narrow advantage of being first to market.

COPAXONE AND CELEBREX HAD SMOOTHER RIDES

The drugs we are following had somewhat less dramatic filings. The NDA for glatiramer acetate was submitted in June 1995, thirteen months after the last observation of a subject in the pivotal trial and five months after the completion of the long-term rat safety study. The FDA responded with an RTF letter because the application provided insufficient information to evaluate the identity, quality, purity, and strength of the drug substance (CDER 1996a). The letter noted the absence of information on how the molecular-weight reference standards were prepared and characterized. The problem was not with the glatiramer acetate but with what it was being compared against. To assure that a drug actually meets specifications, each batch must be tested and compared with a reference standard. For glatiramer acetate, which is not a single compound but a mixture of similar molecules of different sizes, a key specification is the range of molecular sizes and tests that are done comparing the drug with molecular size standards. The NDA failed to provide sufficient information on how the standards themselves were prepared and analyzed.

The RTF letter also noted deficiencies in the documentation for the authorization to the reference standard production documents, inadequate analytic data tables, missing stability data, and lack of a table linking drug substance lot numbers to individual preclinical and clinical studies. The letter also noted that the application did not provide evidence for the ability to scale up production of the drug substance. This would limit production to the pilot scale used for the trials unless and until validated process scaleup information and analytic data for production-size batches were provided and reviewed. Finally, the letter requests data addressing whether antibodies are produced as a result of administering glatiramer acetate and whether such antibodies neutralize drug activity. These deficiencies did not contribute to the RTF.

In fact, the reasons given to support the RTF were less daunting to address than the other deficiencies noted, and the NDA was resubmitted on October 10, 1995 and accepted for filing shortly thereafter.

The NDA for celecoxib, submitted electronically on June 29, 1998, was accepted for filing on August 20, 1998. In those two months, the FDA reviewers made nine requests for information, clarifications, and resubmission of flawed tapes (CDER 1998b).

14

The FDA Review

The next step for a drug is the FDA's review of the new drug application, a process that may seem like a round of dull analyses by faceless technocrats but can turn into a do-or-die situation for a drug. It is worthy of some attention. The FDA's NDA review, in the hands of the assigned team of physicians, scientists, statisticians, and regulatory experts employed by the FDA, is stewarded by a project manager (PM) who assures that all the required meetings are held and that the individual reports that make up the review are completed on time. Considering the scope of the FDA's work in assuring the safety and effectiveness of human drugs, the number of people employed in the effort, about 3,000 in 2003, seems modest. In fiscal year 2001, 101 new drug or biologic submissions were reviewed, plus 168 supplements for new efficacy claims, over 2,000 supplements for changes in manufacturing procedures, and 699 INDs. For fiscal year 2003, the FDA reviewed 113 new applications and 74 resubmissions (FDA 2003c). The reviewers also must respond to requests for meetings from industry, schedule and hold the meetings, prepare the minutes of the meetings, and handle other regulatory administrative tasks (GAO 2002a).

WHO PAYS FOR THE REVIEW?

The 1992 FDA Modernization Act, also called the Prescription Drug Users Fee Act (PDUFA), addressed the number of reviewers and the impact of too few reviewers on the time it took for reviews. The provisions of PDUFA were renewed in 1997 and 2002, as PDUFA II and PDUFA III. PDUFA I was passed in response to concerns raised by the pharmaceutical industry (that engaged congressional attention) charging that drug approvals were languishing at the FDA. David Kessler, then commissioner of the FDA, testified in 1992 that reviews took up to two years because the budget appropriations provided for insufficient medical and other reviewers (U.S. Congress 1992). The proposal that pharmaceutical companies could be charged fees to provide increased funds to hire more reviewers gained credence in 1991 after a review of the FDA's funding and activities (U.S. Senate 1991). The general idea of user fees to provide funds to be added to budget appropriations gained popularity in the Reagan and George H.W. Bush administrations in the late 1980s and early 1990s, and although the pharmaceutical companies were not entirely happy with paying more to the agency that regulated and constrained them, large pharmaceutical companies eventually supported the idea of fees linked to performance goals and legislation with a sunset clause that would allow the user fee program to lapse if the review time did not shorten (Begley 1992).

   PDUFA also introduced two classes of NDAs, standard and priority, assignments that affect the time the review will take. The FDA review team, generally after meeting with the sponsor prior to the submission of the NDA, assigns the NDA to a review class, either standard or priority. If the sponsor proposes and the FDA agrees that the drug has the potential to provide a significant improvement in treatment (the "not me too" issue) because of improved efficacy and/or safety, then the NDA will be assigned priority status. The assignments do not determine how long a particular review will take but are linked to agency performance goals for the length of time for reviews in each class. PDUFA requires the FDA to complete review of 90 percent of priority NDAs in six months, of 90 percent of standard NDAs in twelve months (PDUFA I), and shortened the standard review time to ten months (PDUFA II AND III) (FDA 2003d).

Under PDUFA, three fees are collected: an application fee upon submission of an NDA, an annual establishment fee for each establishment that manufactures a drug in an approved drug application, and an annual product fee for each product in commercial distribution. There are exceptions to the fee requirements, including an exception for orphan drug product applications. The fees are recalculated each year based on the FDA's revenues from the fees, using total fee revenue set by law for each fiscal year. By applying a correction for inflation and projecting the number of each type of application and the workload required for each, the amount of each fee is set. For fiscal 2004, the application fee was $573,500, the establishment fee $226,800, and the product fee $36,080. For the review of human drugs in FY2003, the FDA projected spending of $374,000,000, of which nearly two-thirds was derived from fees from industry (FDA 2003d). For fiscal 2005, the application fee was set at $672,000, establishment fee $262,200, and product fee $41,700 (HHS 2004).

WHO DOES THE REVIEW?

The FDA recruits reviewers on its Web site, posting qualifications and salary ranges (CDER 2003c). Regular employment at CDER in the U.S. Civil Service or Public Health Service Commissioned Corps is limited to U.S. citizens, though permanent resident aliens may be hired through temporary fellowship programs. Physicians must hold a Doctor of Medicine or Osteopathy degree; board certification/eligibility in a medical specialty and experience in conducting clinical trials are "highly desirable." Doctoral-level scientists with at least two years of postdoctoral experience in biology, chemistry, microbiology, pharmacology, pharmacokinetics, toxicology, or epidemiology are sought to evaluate for scientific validity the portions of the INDs and NDAs relevant to their discipline. Positions for project managers require scientific degrees or appropriate course work and, ideally, experience in health care or pharmaceutical industry project management.

Salaries, based on discipline and experience, are standard civil service general services (GS) salaries. For example, the starting salary for a physician ranges from $64,000 to $116,000; for other doctoral-level scientists, the starting salaries range from $51,000 to $61,000; and project manager annual starting salaries start just below $24,000 and go as high as $61,000 (CDER 2003c).

The increase in workload resulting from the performance goals for review times, as well as the increase in the number of meetings required by PDUFA, appears to have taken a toll on the FDA staff. The GAO's review of the effectiveness of the drug user fees provides evidence that the attrition rate of FDA reviewers of all disciplines except chemists is 2–3 times higher than comparable positions at the CDC or NIH (GAO 2002a).

In fact, review times fell between 1987 and 1992, prior to the passage of PDUFA. The standard NDA review time fell each year from 1987 to 1992, from thirty-three months to nineteen months. Only 17 percent were priority submissions, and priority NDA review times were not much shorter, falling from twenty-nine months to seventeen months (GAO 1995). The GAO analysis of the factors influencing the review time suggested that experienced companies' NDAs were reviewed a bit more quickly; delays caused by sponsors accounted for about 20 percent of review times, and nearly half of NDAs were not approved.

The GAO analysis in 2002 of the impact of user fees on drug approval times and withdrawals of approved drugs documented that the increased funding shortened the time for review of a standard NDA to fourteen months in 2001. Priority NDAs had a median review time in 2001 of six months. Review times for biologic drugs were generally longer and fluctuated from twelve to thirty-two months (GAO 2002a).

The assignment to a priority review status, by cutting the time for review, will shorten the time to market, if all goes smoothly. A simple calculation suggests how valuable this may be. A drug with $200 million in annual sales earns nearly $55,000 for each extra day of sales. Six months means real money! Searle requested and received priority review status for the celecoxib NDA. Teva did not seek priority status for copolymer-1.

HOW DOES THE WORK OF THE REVIEW GET DONE?

How the review team goes about its work varies among the divisions of CDER, but glimpses into the process are available from review flowcharts provided by CDER medical officers (CDER 2001b; Honig 2003). CDER is organized into offices, and the Office of New Drugs is itself divided into six drug evaluation offices. Within each drug evaluation office are divisions with responsibility for drug products used in specific types of disorders. For example, the division of anti-inflammatory (such as celecoxib), analgesic, and ophthalmologic drug products is in the Office of Drug Evaluation V. The division of neuropharmacological drug products is within the Office of Drug Evaluation I.

A pre-NDA meeting of the review team with the sponsor is held to organize for the effort. At this meeting, the status of the pivotal trials will be discussed and agreement sought on the studies, preclinical and clinical, to be submitted and the format of the reports and data sets to be submitted. The team will also seek to reach agreement on the key data analyses that will be submitted. Target dates for the submission of preclinical and clinical data will be set. Following this meeting, the review team will go over the IND reviews and comments on the protocols, draft the sections on protocol design, and meet with the FDA statistician to discuss the sponsor's proposed analysis plan and any issues impacting on how the FDA's statistical analyses will be done. Usually the sponsor notifies the FDA one or two days before submission of the NDA, triggering the notification by the project manager of the review team and the division of scientific investigations (DSI). The division of scientific investigations is responsible for auditing clinical investigation sites for pivotal and other critical studies to document that the studies were carried out and documented as described in current good clinical practices. The PM prepares a review Gantt chart, a graphic tool used in managing large projects that presents every task that must be done, how long each takes, who is responsible for each, and the sequence in which the tasks must be done. The PM also consults the schedule of the appropriate advisory committee, panels of outside experts whose advice is sought if the application is for approval of a new drug that works in a novel way, a new or previously approved drug likely to have a major therapeutic impact, or a new drug that presents special problems or issues in evaluation (Mathieu 1994).

The medical reviewer's first task on receipt of the NDA is to review the sponsor's study reports for the pivotal studies to see whether the trials were carried out as agreed to at the end of the Phase 2 meeting and analyzed as discussed at the pre-NDA meeting. The critical analyses for efficacy are identified and discussed with the statistician. Patients excluded from analyses by the sponsor are identified and the reasons for exclusions checked to see if they are sound. Studies where the effects of treatment are marginal, or pivotal trials where the conclusions differ, are identified, and study designs and patient characteristics are evaluated to see whether they account for the marginal or inconsistent effects.

The medical reviewer examines the intent to treat (ITT) analyses. The large Phase 3 trials are done to try to mirror the situation in clinical practice, and the eligibility criteria for the trial may eventually be reflected in the wording of the labeling that describes the patients who may benefit from the drug. An intent to treat (ITT) analysis includes not only patients who completed the course of treatment but also patients withdrawn from the study for any reason other than documented ineligibility. These patients may have been legitimately excluded from the sponsor's analysis of the effects of the entire course of treatment on the condition, but the ITT analyses will provide results more like that anticipated in everyday medical practice, where people do not always take the full course of treatment. The reviewer must be sure that the review provides both kinds of analyses.

To be sure that results are not skewed by an inadvertent or intentional bias at one or more clinical sites, efficacy analyses of pivotal trial data are done for subjects studied by each investigator. Significant safety issues are identified and examined by site. Trends suggesting different efficacy results or adverse event patterns at different sites suggest that site audits may be needed. This information is communicated to the DSI to allow planning for audits.

At a meeting held forty-five days after the filing, when the decision is made whether to accept the filing, the results of the initial look at the submission are discussed and the team's strategy for the review set. The team will decide whether they will request additional case report forms, analyses, and/or patient follow-up information from the sponsor. The medical reviewer and statistician will agree on how the work of the medical review will be divided and establish weekly meetings. Generally the medical reviewer spot checks the validity of the database, sees whether the sponsor's patient exclusions are scientifically appropriate, identifies the sponsor's analyses that should be reanalyzed by the FDA statistician, and specifies the analyses that the FDA will perform for the review.

The statistician will validate the sponsor's analyses, perform the additional analyses, and highlight problems in discussions with the medical reviewer. At the forty-five-day meeting, the PM's Gantt chart for the review, including target dates for completion of the components, are set, though revisions may occur as the monthly meetings of the review team provide updates on progress, problems, and issues.

During the months following the forty-five-day meeting, a study-by-study review is performed for efficacy and an integrated safety review completed. Differences among studies in design, patient characteristics, the frequency of monitoring, and the length of follow-up are identified, and a study-by-study review for safety is performed.

SHOULD OUTSIDE EXPERTS BE BROUGHT INTO THE DECISION?

If the decision has been made that an advisory committee review will be requested, the strategy for the meeting is discussed, focusing on differences between the applicant's and FDA's analyses or conclusions. An important goal of the review is to identify any issues that require the judgment of the members of the advisory committee. A teleconference with the sponsor may be held to discuss areas of disagreement and resolve differences for presentations at the advisory committee meeting. The FDA's briefing package of data and analyses will be sent to the advisory committee, as are the sponsor's package with slides to be presented.

During the review, in addition to the scheduled meetings and teleconferences with the sponsor, a variety of requests for further data, clarification of questions or inconsistencies in any of the components of the application, clinical and preclinical studies, and reports and documentation of the manufacturing and drug and drug product quality control may be made of the sponsor. These will generally be coordinated through the regulatory staff at the company and the FDA.

ONSITE CHECKS ARE THE NORM

To ensure the quality of the data submitted to the FDA for regulatory approval and to protect human subjects of research, the FDA, under the Bioresearch Monitoring Program, carries out site visits and audits of institutional review boards (IRBs) and clinical investigators (FDA 1998b). The review team assigns FDA field offices to carry out study-oriented clinical investigator inspections. The FDA field officer contacts the investigator and sets up a mutually acceptable time for a visit, when the details of the conduct of the study are determined: who did what, how authority was delegated, where study activities (dosing, blood drawing, testing, etc.) were performed, how and where data were collected, how the drug was accounted for, and how the study monitor evaluated the progress of the study and communicated with the clinical investigator. The study data will be audited and compared with the data submitted to the FDA. Records supporting subject eligibility for the protocol may be reviewed, as may follow-up records. The FDA investigator will conduct an exit interview with the clinical investigator to go over the findings and clarify any misunderstandings. A written form of the inspectional observations may be issued, and a report of the site visit is submitted to FDA headquarters.

Following the review of the report, a letter is sent to the clinical investigator that either notes no significant deviations from the regulations, informally identifies deviations from regulations or good investigational practice that may or may not require a response, or provides an official warning of serious deviations that require prompt correction. This information may be communicated to the site IRB and the sponsor. The FDA may institute regulatory or administrative sanctions that could lead to formal disqualification of a clinical investigator. If a clinical investigator notified of serious deficiencies does not provide timely and satisfactory explanations for noncompliance, the FDA will offer the investigator the opportunity for an informal hearing, called a Part 16 hearing (CFR 2003e). For the hearing, the FDA commissioner designates a presiding officer from the Office of Health Affairs, who issues a report on the hearing. All parties in the hearing can review and comment on the report. The commissioner who decides whether the investigator should be disqualified uses all of the documentation. The sponsor is notified of the disqualification. Unless the sponsor can document that the clinical investigator's misconduct did not invalidate the data from that site included in the submission, the data will be eliminated from consideration. If the commissioner concludes that the remaining data are inadequate for the filing, the sponsor may request a Part 16 hearing (CFR 2003e). No such problems were uncovered during the reviews and audits for the NDAs of copolymer-1 and celecoxib.

REVIEW OF THE COPAXONE NDA

The Summary Basis for Approval for Copaxone (copolymer-1) lists nine amendments to the October 1995 submission in response to questions and issues raised by the reviewers. Their nature is unknown.

The medical review of the two copolymer-1 efficacy studies, dated December 5, 1995, first accounted for every subject entered into each study and then presented an analysis of the two pivotal studies. For the NDA, the sponsor reanalyzed the Bornstein study to include the two placebo-group patients who were excluded in the published analysis by Dr. Bornstein because he concluded that for neither of them could the diagnosis of MS be confirmed. The intent to treat analysis confirmed the conclusion of the publication that the proportion of relapse-free patients was significantly greater in the group treated with copolymer-1 than in the control group and that more copolymer-1–treated patients had no or fewer than three relapses during the two-year study.

The larger multicenter trial had a different primary outcome measure, the mean number of relapses over the twenty-four months of the double-blind trial, plus secondary endpoints of the proportion of relapse-free patients, time to first relapse, proportion of progression-free patients, time to progression by at least one point on the standard scale, and change in three scales that measure disability, ability to walk, and total body function. The intent to treat group consisted of the 251 patients randomized, 125 to copolymer-1 and 126 to the placebo. The two-year relapse rate was significantly reduced, by 29 percent, in favor of copolymer-1: 1.19 versus 1.68 relapses per patient for two years, and 0.60 per year versus 0.84 per year. The results for most of the secondary endpoints trended in favor of copolymer-1 but did not reach statistical significance. There was a significant difference favoring copolymer-1 for both the number of patients whose disability scores worsened and the extent of worsening. The medical reviewer concluded that the large trial showed a small, though statistically significant, treatment effect favoring copolymer-1. Yet because the plan agreed to by the FDA and Teva was the performance of a larger, multicenter trial to confirm the Bornstein results (Pinchasi 2003), the issue of the difference in the size of the treatment effect between the two studies was a concern. The reviewers described several differences between the studies in subjects that may have contributed to the difference in the magnitude of the treatment effects. The subjects in the Bornstein group were younger, had been diagnosed more recently, and had a higher two-year prestudy relapse rate. Also, more subjects in the placebo group in the Bornstein study experienced three or more relapses during the trial than the placebo group in the larger study. These issues were noted, but based on confidence in the statistical confirmation of the primary endpoint, the conclusion of the reviewer was that "based on these two studies, copolymer-1 appears to reduce the frequency of exacerbations in patients with exacerbating-remitting multiple sclerosis" (CDER 1996a).

The safety review of the copolymer-1 NDA started when the medical review was complete, required several months to complete, and was revised a few months before the advisory committee meeting in September 1996 (CDER 1996a). The copolymer-1 safety review is highly structured, beginning with a narrative introduction of the discovery and testing of copolymer-1 as well as comments on the proposed labeling. Descriptions of sources within the NDA used for the safety review (the submitted integrated safety summary, individual study summaries, data listing, case report forms, patient narrative, etc.) and methods for the review are described. Regarding the quality of the submission, the reviewer found that the submission met the criteria for filing and proceeded to describe deficiencies in the content, organization, and indexing of the submission while noting that the coding of adverse events, adherence to protocol design, and surveillance for adverse events were acceptable. The audit of the CRFs and patient narratives found no misreporting or contradictions. Problems with the indexing and cross referencing in the database and terse patient narratives did present problems for the reviewer in assessing adverse events that needed further explanation. The total number of people receiving copolymer-1 was 852 people with MS in 11 controlled and uncontrolled clinical trials. Of these, 779 had relapsing-remitting MS and 73 chronic MS. One measure of the human exposure to a drug is patient years, found by totaling the length of time the drug was given to each patient. That information was available for 736 patients and amounted to 1,092.4 patient years. Overall, the subjects in relapsing-remitting MS trials were representative of those likely to receive the drug: two-thirds female and of mean age 36.8 years. Eighty-six percent of 426 subjects receiving copolymer-1 were recorded as Caucasian. In the NDA, seven deaths were reported, two in relapsing-remitting groups and five in the chronic MS group; all were receiving copolymer-1. Two patients who died were described as having experienced the systemic reaction (a feeling of tightness in the chest, shortness of breath, and anxiety) weeks or months prior to death. The reviewer finds the records insufficient to allow any conclusion as to whether the deaths were caused by the study medication, specifically by the systemic reaction. The incidence of other serious adverse events was too low to establish cause. Approximately 23.7 percent of patients receiving copolymer-1 withdrew from a study. Twenty-one percent of the relapsing-remitting subjects receiving copolymer-1 in the two placebo-controlled trials withdrew, most because of adverse events, specifically injection site reaction. No more than four patients dropped out because of symptoms of the systemic reaction. A table of the adverse events that occurred in 5 percent or more of the subjects receiving copolymer-1 or the placebo in the controlled studies, a table that is incorporated into the package insert, was provided. Adverse events are discussed according to the body systems involved using an early adverse event mapping dictionary, COSTART. Tremor, confusion, and agitation (neurology) were more common with copolymer-1 than a placebo, as were chest pain and faintness and swelling of the face caused by fluid retention. Serious adverse events for the entire database were slightly more frequent in the copolymer-treated subjects (4.5 percent) than in the placebo group (3.4 percent).

The safety review of copolymer-1 discusses at length the issue of whether copolymer-1 causes an immunologic or allergic response. In the large multicenter trial, subjects were tested every 3 months for the development of antibodies to copolymer-1; 80 percent of copolymer-1–treated patients had peak levels of antibody greater than 150 percent over baseline within 3–6 months of exposure, dropping to 50 percent over baseline in most. Levels in 80 percent of placebo subjects were less than 50 percent over baseline, and antibody was detected at random time points. The reviewer discusses animal and laboratory tests of the ability of copolymer-1 to cause an immune or allergic response and opines that the "systemic reaction" symptoms are consistent with a generalized allergic drug reaction.

A concern raised, but not elaborated on, by the safety reviewer, based on the animal toxicology studies, is whether antibodies bind to copolymer-1, creating complexes that deposit in the kidney, causing damage to the kidney. The systemic reaction is also discussed at length, with a focus on whether it was inadequately defined and therefore underreported. The lack of understanding of the cause and nature of the systemic reaction is clearly a concern to the reviewer, as is the incomplete explanation for the injection site reactions.

In his conclusion, the safety reviewer states that from a safety standpoint the NDA for copolymer-1 is approvable and then recommends eight issues to be "explored" by the sponsor: clarification of the time course of the drug's absorption into the body and its elimination, the relation of the effect of the drug to dose, a study to rule out in humans a type of autoimmune disease seen in some of the monkey safety studies, a study to determine what causes the injection site reaction, a study to more fully understand the chest pain seen in some subjects to rule out loss of blood flow to the heart, a study to better describe and understand the systemic reaction, surveillance after approval (postmarketing surveillance) for evidence of inflammation of blood vessels and a variety of conditions caused by immune responses to drugs, and a discussion between the sponsor and the FDA to develop an improved definition of the systemic reaction.

Thus the medical and safety reviewers conclude that the NDA is approvable and raise a series of issues for the advisory committee.

REVIEW OF THE CELEBREX NDA

The review of the NDA for celecoxib, because the sponsor sought approval for three indications—management of pain, rheumatoid arthritis, and osteoarthritis—and included reports on fifty-one separate clinical trials, is a longer and more extensive document (CDER 1998b). Yet the indications were more familiar to the FDA because a number of drugs had been tested and approved for each indication in the recent past. The novelty of this application rested on the claim that celecoxib is a selective COX-2 inhibitor. The implications for efficacy and safety of the suggested selectivity, central to the sponsor's claims for novelty, were unknown.

The medical review of celecoxib for the pain studies first addressed the four single-dose active drug and placebo-controlled studies in post-surgical dental pain and concluded that there was a graded response to different amounts of celecoxib and that the pain relief with celecoxib lasted longer but the active controls (aspirin, naproxen sodium, or ibuprofen) provided faster and more complete pain relief. The review concluded that the results of the three multiple-dose pain trials in postorthopedic surgery pain did not provide consistent evidence that celecoxib was better than a placebo, possibly because this setting requires the higher degree of pain relief provided by opiates.

The medical review of celecoxib for osteoarthritis dealt with ten studies: five twelve- or six-week pivotal trials, two short placebo-controlled supportive dose-ranging studies, one active drug controlled six-week supportive study performed outside the United States, and two active controlled supportive studies to assess the incidence of stomach ulcers. A total of 7,239 patients were enrolled in these trials,

4,021 of whom received celecoxib. Based on intent to treat (ITT) analyses, the reviewer concluded that celecoxib from 100 to 200 mg twice a day is consistently effective in treating the signs and symptoms of osteoarthritis compared with a placebo and is comparable to 500 mg of naproxen twice a day. Patients still had some symptoms; the effects of celecoxib were durable for twelve weeks, though there may be some waning of the response over time.

For rheumatoid arthritis, the medical review addressed seven studies, two placebo and active drug controlled pivotal trials of twelve weeks, one four-week placebo controlled pilot supportive trial, and four active drug controlled supportive trials. Results are analyzed for a total of 3,672 rheumatoid arthritis patients, 2,098 of whom received celecoxib. The reviewer concluded that celecoxib efficacy in treatment of the signs and symptoms of rheumatoid arthritis was demonstrated. ITT analyses indicated that celecoxib from 100 to 400 mg twice a day was effective when compared with a placebo and comparable to naproxen at 500 mg twice a day. The higher doses, 200 and 400 mg, of celecoxib have comparable efficacy and are frequently more effective than 100 mg twice a day.

The safety review for celecoxib first attempted to account for all individuals in the submission exposed to celecoxib, noting that for any submission, totaling from the tables of each trial sometimes leads to small discrepancies because patients may be represented more than once. The number settled on for the celecoxib NDA review was 7,851; 1,992 patients received the proposed dose, 200 mg twice a day, for 92 days or more, up to 450 days. Exposure data were also drawn from a large open label (nonblinded) safety and efficacy study ongoing at the time of the submission. A safety update for 5,629 subjects with 120 days of exposure plus subjects in six controlled clinical trials not included in the NDA and the open label study brings the total to 10,740 subjects exposed to celecoxib. Patients in the OA trials were generally older than 40 years, with a mean age of 58; the mean age of the RA subjects was 59.5 years. Females predominated, and 82 percent of subjects were Caucasian. The subjects in the Phase 1 and pain trials were less homogeneous and tended to be younger.

The safety analysis of celecoxib dealt not only with serious and nonserious adverse effects but also with specific safety issues based on what was known about COX inhibitors in general and what was known about COX-2 and celecoxib. The safety review is thus not just a survey of adverse events but a science-based examination of changes in suspected tissue targets for the drug under review and the general class of drugs. NSAIDs, nonspecific COX inhibitors, are known to inhibit the function of platelets (the small cell fragments in the blood that are important in blood clotting), to cause reductions in kidney function, and to cause damage to the lining of the stomach and small intestine. Changes in platelet function can have negative effects on blood vessels and the heart, so cardiovascular effects were specifically analyzed. Because celecoxib is broken down in the liver, the reviewer carefully looked for evidence of changes in liver function. Because it appeared that COX-2 might be present in the pancreas, metabolic safety was analyzed. Because patients with OA or RA are generally older, specific effects in older patients were also analyzed.

To assess the effects on platelet function, the sponsor reported five studies in a total of 166 healthy young and elderly individuals in which platelet function was tested, including the tendency of the blood platelets to clump together and the blood level of a chemical called thromboxane that is required for blood clotting. In some of the studies, the ability of the blood to clot was tested directly by measuring how long it takes for the bleeding to stop after a small incision is made in the skin. These studies were controlled with both a placebo and active drug; several included celecoxib at doses higher than the recommended dose of 200 mg twice a day. The results showed that compared with a placebo, celecoxib, unlike standard NSAIDs, did not affect platelet function. Also, adverse events caused by bleeding were not associated with celecoxib use.

No obvious patterns of changes of blood and urine glucose or metabolic adverse events were associated with celecoxib. No changes in liver enzymes or adverse events caused by liver or bile duct disease were associated with exposure to celecoxib. The reviewer noted that the pharmacokinetic studies showed a slower removal of the drug from the body in subjects with mild liver disease and the absence of studies in subjects with severe liver disease, suggesting that patients with severe liver disease should not be given celecoxib until further studies are done.

The assessment of effects on the kidney was more complex. There was a pattern of changes in laboratory tests and adverse events related to kidney function that was different from the placebo and similar to NSAIDs. There was no strong evidence of celecoxib causing serious kidney disease, though the reviewer indicated that to detect a link would require a larger database and more subjects.

The safety of celecoxib for the stomach and small intestine was the core of Searle's strategy for celecoxib, based on the hypothesis that the selectivity for COX-2 will allow celecoxib to provide anti-inflammatory effects and pain relief without causing damage to the cells lining the stomach and small intestine. Two measures were used to test the hypothesis: (1) the development of ulcers and visible erosions of the stomach and the adjoining part of the small intestine; and (2) the occurrence of clinically significant adverse events involving these tissues. Six trials in over 4,700 healthy subjects and patients with RA or OA used endoscopy, the insertion of a tube into the stomach, to see whether there were signs of leaky blood vessels, erosions of tissue, or ulcers. These studies included placebo and NSAID controls. The reviewer comments that celecoxib was generally significantly different from the NSAIDs in that it was associated with fewer ulcers, with the exception of diclofenac. The reviewer also writes that the incidence of ulcers with celecoxib was similar, though not equivalent to, that seen with the placebo. Searle had assigned the process of comparing clinically significant upper gastrointestinal (UGI) events, such as gastric bleeding, perforation, or obstruction in the outlet of the stomach, with an independent gastrointestinal consultants committee of three gastroenterologists, who without knowing what drug the subject was receiving reviewed the case summaries of potentially clinically significant UGI events provided by the site investigators, who were instructed to report such events immediately. A total of 170 potential cases were reported; the committee judged only 18 as meeting the criteria of a definite UGI event. Eleven occurred in the controlled arthritis trials: nine in subjects receiving an NSAID, two in subjects receiving celecoxib, and none in placebo subjects. The reviewer comments that the small number of events does not allow any valid statistical inference to be drawn, though the trend supports that for the occurrence of clinically significant UGI events celecoxib is different from the NSAIDs with which it was compared and not the same as the placebo. When the incidence of clinically significant UGI events in the open label trial and the 120-day safety update was analyzed, the annual rate was similar to the annual rate for ulcers found with endoscopy tracking in the six controlled trials, though the reviewer cautions that the number of clinically significant UGI events is too small to provide statistical evidence that endoscopy predicts clinically significant UGI events. The question of whether celecoxib will "get" the labeling Searle seeks, that it is less likely than NSAIDs to cause damage to the lining of the stomach and small intestine, is unanswered in the safety review narrative.

The medical and safety reviews of the two drugs have accomplished two things: provided careful analyses of the efficacy and safety of each drug, and in the highlighting of issues set the stage for the respective advisory committees.

15

What Do Outside Experts Think? The Advisory Committee Meeting and FDA Approval

The time has arrived. The members of the advisory committee have finished their hotel bacon and eggs, made the necessary phone calls to the office to make sure there were no wildfires that needed dousing, and assembled in the unadorned conference room in the familiar and tired office building in Bethesda. The audience is getting settled, including the financial analysts in the back rows with cell phones at the ready so they can share the nuances of the discussion with their anxious colleagues. The company and FDA teams are ready with their slide carousels (or, more recently, loaded laptops or memory sticks) and importantly the indexes for the hundreds of slides, so that they can quickly respond with data to the question or concern of a committee member. The company team, their business suit attire as ritualized as a toreador's red cape, will make their presentations, reviewing the study results and arguments for approval with the most expansive label, the description of exactly for whom the drug is approved. Each team member may suspect that their professional future hangs on their poise today. Each is confident that they know their stuff. That is not the question. The issue is whether they can provide the committee members with the intellectual comfort food needed not only for approval but also for the broadest possible label, providing the large market that the company wants for the drug and, in the case of some companies, needs to survive. The FDA reviewers will present their analyses, their concerns embodied in the questions that were sent to the committee members and the sponsor just a few weeks earlier. The committee members earlier had received the company's briefing package of slides to be presented. There also may be brief statements by a chorus of patients, patient representatives, and spokespersons for public interest groups, each with something to say pro or con about the drug under review. The committee will then move through the FDA's questions in a discussion managed by the committee chair and strained by the competing interests of allowing a full and complete airing of the data, questions, concerns, and arguments and actually reaching a vote on each question. Working through the FDA's questions, the committee dutifully decides on approval and on the draft of the label, which will be negotiated with more precision later by the company and the FDA. It will be a long day, a long performance of scientific debate with several acts and a high likelihood of intense though generally subdued drama, the stage set unchanging during the performance but the emotional score rising and falling with an unpredictable rhythm. No doubt this is theater, with exposition, complications, climax, resolution, and generally conclusion, and sometimes, though rarely, catastrophe.

WHO ARE THESE ADVISORS, AND WHY ARE THEY SITTING IN JUDGMENT?

The penultimate step for many but not all drugs is pure theater, the theater of an FDA advisory committee meeting. The FDA's advisory committees, composed of expert physicians and scientists outside of the FDA, provide recommendations and opinions on the safety and efficacy of applications to market new drugs and on FDA policies; CDER currently has sixteen advisory committees, which cover agents for specific medical conditions such as arthritis, skin and eye disease, heart and kidney diseases, viral infections, and other disorders, plus committees whose charters cover nonprescription drugs, drug safety and risk management, and pharmaceutical science. The director of each division, in consultation with senior medical officers, makes the decision to submit an application for an advisory committee review.

The executive secretary of the appropriate committee is consulted and the application placed on the agenda of an upcoming meeting.

The FDA's use of outside advisors began after the disclosure of the thalidomide birth defects triggered passage of the 1962 amendments to the Food, Drug, and Cosmetic Act (IOM 1992). Ad hoc committees were established to provide advice on birth defects and to help in the review of the first birth control pill. Standing committees were established in 1967, and with some interruptions for bureaucratic tinkering, the FDA advisory committees, functioning under the rules of the 1972 Federal Advisory Committee Act (FACA), the 1976 Government in Sunshine Act, and the 1978 Ethics in Government Act, have come to play a major role in the approval of new drugs. The recruitment and management of outside advisors began with help from an outside agency, the National Academy of Sciences. The 1962 amendments mandated that the FDA not only review the safety and efficacy of new drugs but also review existing drugs for efficacy. To accomplish the enormous and potentially politically unpopular task of reviewing the nearly 3,000 different approved and nonapproved drugs on the market in 1962, the commissioner of the FDA, James Goddard, in 1966 convinced the National Academy of Sciences to establish, through its National Research Council, a series of advisory panels. Each of the thirty panels of six members reviewed drugs for a particular purpose; for example, antibiotics to treat infections, drugs for heart disease, and so forth. The drugs were categorized as effective, probably effective, possibly effective, or ineffective based on reports submitted by the companies, the scientific literature, FDA files, and the experience and opinions of the panel members. When the report of the NAS-NRC was submitted in 1969, recommendations on over 2,800 drugs were provided, including recommendations that some drug approvals be withdrawn (National Research Council 1969). If the FDA decided to withdraw approval, the company was afforded the opportunity to appeal, and in some cases litigation followed (IOM 1992). Although the NAS-NRC panel primarily considered prescription drugs, they also reviewed over 400 over-the-counter (OTC) drugs, but the backlog of thousands of OTC drugs approved between 1938 and 1962 required the establishment of a system of advisory panels for OTC drugs and a review system for new OTC drugs (IOM 1992).

CDER's advisory committees, one or more advising each product division, function under the FDA regulations to provide advice on drug approvals and on drug development guidelines (CFR 2003f). Although the formal intent of the advisory committee system is to have the highest-quality review of new drugs, the committees also provide credibility to the decisions, some of which may be controversial (IOM 1992). The agendas of committee meetings begin with introductions of those participating, and the committee's executive secretary reviews the conflict-of-interest information and waivers for members of the committee and all those participating in the discussion and the voting.

How the FDA handles potential conflict of interest or perceived conflict of interest for FDA advisory committee members has triggered some controversy, dispute, and litigation (Golde 2002). The members and consultants to an advisory committee are special government employees, and the terms of the criminal code 18 USC 208 prohibit a government employee from participating "personally and substantially in a particular matter" in which he or she has a financial interest (FDA 2000b; U.S. Code 2001). The code does provide for waivers if the FDA commissioner determines that the interest is insubstantial or if the "need for the individual services outweighs the potential for conflict of interest" (U.S. Code 2001). 18 USC 208 2(b) also provides for a waiver if the Office of Governmental Ethics (OGE) determines that the potential for conflict of interest is remote (U.S. Code 2001). The 1997 Food and Drug Modernization Act expanded the prohibition to include financial benefit to the member's immediate family. The process used for vetting nominated committee members includes collecting detailed information on the nominee's financial interests to determine the extent of financial ties to FDA-regulated firms. The ties may be so many and so large that the nomination is withdrawn. The FDA solicits annual updates of the financial ties of committee members to FDA-regulated firms. The executive secretary and the Office of Special Government Employees (SGE) review the potential conflicts of interest of committee members and proposed guest participants for the submission going before a committee. If a potential conflict is identified, the executive secretary, the division director, the office director, and the SGE program officer decide whether to seek a waiver from the FDA commissioner. The commissioner holds authority to grant waivers under 18 USC 208 2(a) and 2(c) and the OGE under 18 USC 208 2(b).

A series of controversies about real or perceived conflicts of interest in 1991 led FDA Commissioner Kessler to request a review of the oversight and legal review of waiver practices by the National Academy of Science's Institute of Medicine, a review that led to a series of legislative and procedural changes that provide for enhanced legal review of potential conflicts and the process of requesting and approving waivers (IOM 1992).

THE ADVISORY COMMITTEE MEETING
FOR THE COPAXONE APPLICATION

The Peripheral and Central Nervous System Drugs Advisory Committee met on September 19, 1996, to consider the application for Copaxone (CDER 1996). Eight committee members were present, including Sid Gilman, M.D. (chair of the committee), Orlando Carter III, M.D., and David Drachman, M.D., each a senior academic neurologist with extensive clinical and research experience but with no publications that suggested a research interest in MS. Other committee members participating were Claudia Kawas, M.D., Associate Professor of Neurology at the Johns Hopkins School of Medicine; Chris Glennings, Ph.D., Assistant Professor of Biostatistics at the Medical College of Virginia; Zaven S. Khachaturian, Ph.D., a former FDA staff member with expertise in neurobiology, who held a position with the Ronald and Nancy Reagan Research Institute of the Alzheimer's Association; Ellyn C. Phillips, the president of the Philadelphia Chapter of the Amyotrophic Lateral Sclerosis Association, a consumer representative; and Patricia Coyle, M.D., Professor of Neurology at the State University of New York at Stony Brook, an experienced MS clinician and researcher.

Seven patient participants were present, two under the auspices of the National Multiple Sclerosis Society. The FDA representatives were Paul Leber, M.D., and Russ Katz, M.D., respectively the director and deputy director of the Division of the Neuropharmacological Drugs, and their supervisor, Robert Temple, M.D., Director of the

Office of Drug Evaluation I. The Teva Pharmaceuticals team included Senior Vice President Carole S. Ben-Maimon, M.D.; Kenneth Johnson, M.D., Professor and Chair of the Department of Neurology at the University of Maryland Medical Center, Baltimore; and Jerry Wolinsky, M.D., Professor of Neurology and Director of the Multiple Sclerosis Research Group at the University of Texas Health Science Center. Dr. Johnson was the lead investigator, and Dr. Wolinsky participated in the multicenter trial under review. Dr. Ben-Maimon, who led the clinical development of Copaxone at Teva USA, had trained as an internist and worked in the pharmaceutical industry since 1991, first at Wyeth and now at Teva Pharmaceuticals.

Ermona McGoowin, the executive secretary, in describing the results of the FDA review of the financial statements of the participants, noted that though they presented no potential conflict of interest at this meeting, Dr. Coyle and her employer, the State University of New York, held interests that could create an appearance of a conflict of interest, and so Dr. Coyle would participate in the discussion but would have no voting privileges. Remember that she is the only committee member with MS research credentials.

Dr. Leber, for the FDA, set the stage with the two questions posed to the committee:

1. Are the two studies "adequate and well-controlled clinical investigations and does each provide evidence that would allow an expert knowledgeable and experienced in the management of patients with MS to conclude that Copaxone is an effective treatment for MS?"
2. Has the sponsor provided evidence that Copaxone is safe when used in the treatment of multiple sclerosis? (CDER 1996b)

Dr. Leber emphasized that the FDA brings to the advisory committees the decisions that are difficult and may involve judgment and sentiment. He said that the FDA is asking whether experts in the area would conclude from the evidence that the drug will have the efficacy the sponsors claim and that the drug was safe in the setting in which it will be used.

Dr. Katz's presentation, although concluding that there were two adequate and well-controlled studies, raised the efficacy and safety issues of concern to the FDA, the themes for the subsequent discussion by the committee. His initial remarks created an unusual dramatic element in this review—Dr. Bornstein, who had died in 1995. In the normal course of a review, the FDA raises questions about the performance of the pivotal studies. The FDA had questions and issues about details of the planning and conduct of the Bornstein pivotal trial that were not included in the submitted NDA and would be difficult to answer in his absence, though Teva was credited for having found several key documents. The unanswered efficacy questions had to do with the proposed size of the trial and the statistical significance of the results of intent to treat analysis related to the two subjects dropped because of psychiatric reasons. The original Bornstein protocol that Teva found and provided to the FDA called for forty patients, not fifty. A second document they provided called for fifty enrollees. This raised a statistical question about the trial analysis. If an interim analysis was done at the forty-patient stage and the decision was made to add more subjects, the statistical analysis must be corrected for "taking multiple looks." Dr. Katz also noted that the primary outcome used for the Bornstein trials, the number of subjects without exacerbations, when applied to the second trial did not show a statistical difference favoring Copaxone.

Dr. Katz raised three safety issues that were discussed several times during the meeting, the chest pain seen in 26 percent of Copaxone subjects and 10 percent of control subjects in the large multicenter trial; the poorly understood systemic reaction; and the lack of results of long-term animal carcinogenicity (tumor development) studies.

Three members of the Teva team built their case for approval with detailed presentations. Dr. Johnson presented a review of the biology and clinical course of MS and put in context the clinical outcome measures used in the trials. Dr. Ben-Maimon provided the preclinical safety studies, the efficacy data from the two controlled trials, and the safety data from the controlled trials and the open label trials, including the treatment IND approved in 1993. Dr. Wolinsky presented a clinical assessment of the use of Copaxone in patients with relapsing MS.

Dr. Ben-Maimon led off. During her presentation, Dr. Gilman asked a series of questions about the description, incidence, and details of the systemic reaction. Dr. Ben-Maimon and, at her request, Dr. Johnson answered these questions, explaining that the systemic reaction was never observed or monitored by physicians, though there was no doubt that a similar set of symptoms were credibly reported by the patients. There was no evidence from the early studies of Bornstein that the incidence was related to dose. In all the controlled studies, the severity and frequency of the systemic reaction did not increase over time, suggesting that it was not an allergic reaction. Dr. Ben-Maimon was very clear: The cause of the systemic reaction was not understood, and there were no lasting effects. Dr. Gilman began a line of questioning, picked up and expanded by Dr. Leber, as to whether complexes of antibody and Copaxone may develop and deposit in the kidneys. At Dr. Ben-Maimon's request, Dr. Frances Mielach, a Teva consultant who had reviewed the animal safety studies, reported that this was seen at a very low level in animals dosed with large amounts of Copaxone. The discussion moved on to laboratory studies to see whether Copaxone causes genetic alterations or tumors. Although there were no results suggesting that Copaxone increased the rate of genetic changes in laboratory studies or caused tumors in animals, the absence of completed long-term carcinogenicity studies did trigger a discussion in which it became clear that Teva had been led to believe that long-term carcinogenicity studies would not be required. Who gave this advice was not clear, but the FDA staff seemed to feel that, though not ideal, the review could go forward. Dr. Katz noted that the FDA did accept the filing.

The issues of the statistical impact of the two patients dropped from the Bornstein study and the forty or fifty enrollment plan/interim analysis were both discussed several times during the meeting and resolved with input from Dr. Ben-Maimon and Dr. Aaron Miller, who had been working with Dr. Bornstein, serving as the examining neurologist on the study. At times, the discussion seemed like an archaeological exercise to uncover and reconstruct what Dr. Bornstein planned and wrote in rejected and revised and subsequently approved NIH grant applications in the late 1970s and early 1980s that funded the study and what was in the protocol the IRB approved.

The conclusion of this lengthy discussion was that these issues did not detract significantly from the power of the Bornstein study.

Dr. Ben-Maimon's presentation of the multicenter trial stimulated a discussion of whether the data supported an effect of Copaxone on the worsening of the condition of the patients, a discussion that focused on statistics, the power of the trial to detect an effect, and the inadequacies of the scale used to track the progress of the disease. FDA staff, Drs. Ben-Maimon, Johnson, and Wolinsky, and committee members, but not Dr. Coyle, participated in the discussion, with the FDA having the final comment that though a trend favoring Copaxone was seen in both studies, it did not reach statistical significance.

The use of magnetic resonance imaging (MRI) was introduced and discussed by Dr. Wolinsky. At that time, MRI was just beginning to be used to study what was happening in the brains of people with MS. It was not raised as contributing to the question of efficacy.

During Dr. Ben-Maimon's presentation of the safety data, the questions focused on the details of the seven deaths in the Copaxone clinical program; none were scored by the reporting physicians as having been related to Copaxone and all were in open label studies. The presentation and discussion of adverse events focused in detail on injection site reactions, chest pain, and the systemic reaction. To Dr. Coyle's question of whether there was a relationship between the site of injection or the patient's size and the incidence of chest pain or the systemic reaction, Dr. Ben-Maimon responded that they looked for such relationships and also looked at the data to see whether there was a link between Copaxone and other medicines the patients were taking or to the level of antibodies and found none. The mechanism of the systemic reaction was again discussed at length, with no resolution.

After Dr. Ben-Maimon presented Teva's conclusions about the efficacy and safety of Copaxone in reducing relapse frequency and changes in disability in patients with relapsing MS, Dr. Wolinsky "highlighted the medical management issues raised by the results of the Copaxone trials," leading off with a tabulation of the financial impact of MS (CDER 1996b). He noted the benefits at all levels of disability and the fact that the drug was very well tolerated. The injection site reactions were mild and the systemic reaction infrequent, unpredictable, and self-limited. Dr. Wolinksy closed by noting that for physicians who treat this serious condition with no prevention or cure, Copaxone is a novel clinical option. He responded to questions about how Copaxone would fit into the choices available to MS patients and their physicians, suggesting that it would be a good choice for a number of patients, particularly those who are continuing to have relapses at high frequency or significant side effects on one of the interferons, or for recently diagnosed patients concerned about interferon side effects or how long a drug will be effective. At the time, both physicians and patients were becoming concerned that neutralizing antibodies to the interferons developed frequently, compromising efficacy.

After the lunch break, the discussion began with Dr. Gilman's presentation of his view on the data, that two well-performed, well-controlled trials showed a statistically significant, though mild, benefit. Dr. Coyle agreed, and after an extended reprise of earlier exchanges with questions from several committee members on the statistical issues, the uncertainties about what precisely the drug is and how it works, and the efficacy outcomes supported by the data, she was asked if this would be a useful drug. She said yes and that she felt in the end it would become apparent that reducing relapses would affect disability and the disease process.

The issue of safety was addressed next; Dr. Gilman, in his role as chairman, summarized the results, again voicing concern about the systemic reaction, the lack of long-term animal studies, and the question of whether antibody-Copaxone complexes will develop and have a detrimental effect on the kidney. He characterized all of these problems as minor. Dr. Coyle agreed, noting that though the systemic reaction and chest pain issue were puzzling, there was no morbidity attributed to the drug in the nearly 1,000 patients treated. Then the discussion revisited the absence of results of long-term animal carcinogenicity studies. Dr. Temple assured the group that the decision to approve, if reached, might be reversed if pending results came back with problems.

The open public hearing followed, in which six people with MS spoke or had their written statements read, each describing their experience with Copaxone and why they urged the committee to approve the drug. A spokesperson for the National Organization for

Rare Disorders also urged approval, describing Teva's provision of the drug without cost to indigent patients under their treatment IND.

Throughout the meeting, the tone of the questions and answers, as judged from the transcript, was professional and collegial. With the exception of Dr. Coyle, who could not vote, the committee members were not "MS-ologists" and, as expected, the expertise of the academic investigators in the trials, Drs. Johnson and Wolinsky, seemed to carry substantial weight.

The committee voted unanimously to approve with respect to efficacy, after assurance that the label would be limited to relapse rate and not say anything about disease progression, and also voted for approval unanimously with one abstention on safety. After comments by Dr. Temple indicating that they would be talking to the company about issues that "need pursuit after approval," and announcements about upcoming committee meetings, the meeting adjourned at 2:54 P.M. (CDER 1996b). Copaxone was approved for sale on December 12, 1996.

THE ADVISORY COMMITTEE MEETING FOR THE CELEBREX NDA

The meeting of the Arthritis Advisory Committee on December 1, 1998, to review the Celecoxib (celecoxib) NDA was a more dramatic affair than the meeting for Copaxone, involving more presentations, more data, and more financial and business significance. Although the discovery and development of Copaxone were based on a working hypothesis about MS, in the end the issues were empirical because how the drug had beneficial and adverse effects was and remains unknown. The clinical trials were not a test of a biologic hypothesis, and the number of likely users of Copaxone was relatively small (fewer than 300,000, and probably much fewer). Celecoxib was selected on the basis of its effect on a set of enzymes, the COX enzymes, and its development was not only the test of a drug but also the test of a specific hypothesis, that selection for COX-2 inhibition would provide an effective and *safer* treatment for common conditions such as arthritis and pain. It also was a test of a general hypothesis that drugs chosen for development based on their selectivity in test-tube assays would have the expected clinical benefit. The financial appeal of a safer, effective drug to treat arthritis and pain was not small. According to the Arthritis Foundation, over 20 million Americans have osteoarthritis; 2 million have rheumatoid arthritis (Arthritis Foundation 2004). As for pain, we all experience it now and again, and the market for prescription NSAIDs, including their use in arthritis, was in 1997 in the billions of dollars (Pain Management 1997).

Since 1965, fifteen prescription NSAIDs had been approved, and Searle's Daypro was scheduled to lose exclusivity in 2001. The toxicity of both the old (aspirin) and the newer, more potent NSAIDs had become a growing concern and led in 1986 to the addition of a template to the labels of NSAIDs that described the gastrointestinal (GI) toxicity (Bjorkman 1998). The language of the GI adverse effects paragraph alerted patient and physician that serious upper gastrointestinal toxicity could occur with the use of these drugs, with or without warning symptoms. Patients with a prior ulcer or bleeding, older patients, those taking other medications, and those in poor health were described as being at elevated risk. The risks of serious damage continued with continued use of the NSAIDs and could be minimized by using the lowest dose for the shortest time (Witter 2001; American Journal of Medicine 1988; Insel 1996). The problems described in the label succeeded in giving pause to physicians considering the use of these drugs, hesitancy that was believed to slow sales as well as protect patients. And with no evidence that any one of the fifteen available NSAIDs was more effective or less toxic, the market potential for an effective and safer NSAID was very large. The goal was approval with the least onerous safety warning; ideally the removal of the NSAID GI warning. The lead player in the day's drama was Dr. Philip Needleman, who had championed the idea that a COX-2 selective inhibitor would be a safer alternative and had led the discovery, selection, and entire development of celecoxib. This was his baby.

The FDA and the Arthritis Advisory Committee had previously outlined the day's script at a meeting in March 1998 when the committee, bolstered by a gastroenterologist (an expert in diseases of the digestive system) and a nephrologist (a kidney disease expert) debated and discussed the safety issues of COX-2 inhibitors, providing input to the FDA on what studies and results would be needed to determine the relative safety of COX-2 inhibitors not only for the stomach and intestines but also for the liver, bones, kidneys, and in general to a pregnant woman and her fetus (CDER 1998a). At this meeting, just three months before the Celecoxib NDA submission, a gastroenterologist from Searle made a brief presentation on the potential for improved safety, particularly for the GI tract, of selective COX-2 inhibition, and as a member of the audience, Dr. Needleman argued for a clinical strategy showing, in placebo and active controlled studies, efficacy comparable to and safety better than the active drug, the development strategy used by Searle (CDER 1998a). At this meeting, a SmithKline Beecham gastroenterologist presented arguments that predicting improved safety based on COX-2 selectivity was overly simplistic; safety needed to be demonstrated by thorough clinical studies.

The attention of the business community and media was also drawn to the Celecoxib NDA review. As early as 1996, the *Wall Street Journal* wrote of the potential of COX-2 inhibitors (Langreth 1996). They were in some cases referred to as superaspirins, as in the June 15, 1998, *New Yorker* piece by a renowned Harvard physician, Jerome Groopman, M.D., lauding the anecdotal success and potential benefits of celecoxib, then called Celebra, and in a September *Time* magazine article highlighting the potential of COX-2 inhibitors and biotechnology-based drugs in development for arthritis (Groopman 1998; Lemonick 1998).

At the December celecoxib review meeting, five voting consultants joined the six members of the Arthritis Advisory Committee (CDER 1998d). Three of the consultants were professors of medicine and rheumatology, one an epidemiologist specializing in arthritis, and one a kidney specialist. Two nonvoting members, a gastroenterologist and a pediatric rheumatologist, also participated.

The FDA posed eleven questions to the committee. These ranged from whether celecoxib should be approved for the indications of the treatment of the signs and symptoms of OA and RA to the question central to the marketing of the drug, whether, based on the reduction in ulcers seen with endoscopy, qualifications to the NSAID GI warning template should be made. The FDA also asked whether the committee agreed with its conclusion that additional data were needed to support the acute pain indication. Several questions addressed whether specific recommendations be made regarding the dose to be used in elderly patients and patients with liver disease. Finally, the committee was asked for recommendations for any Phase 4 studies, studies to be required after approval.

After the introductions, the executive secretary noted that waivers had been granted to three members of the Arthritis Advisory Committee, including the chairman, and described the financial interest that might be perceived as presenting a conflict for the two invited nonvoting guest speakers.

The Searle presentation of celecoxib's development, introduced by Dr. Needleman, covered the background rationale and discovery of the drug, the results of the preclinical studies, the data on how humans take up, distribute, and dispose of the drug, the results of the clinical trials for efficacy in the three proposed indications—osteoarthritis, rheumatoid arthritis, and pain—the human safety data, and lastly the ways in which the sponsor believes celecoxib is different from NSAIDs. The basic science presentations, detailed and focused on demonstrating that for arthritis and pain COX-1 was not relevant and that celecoxib was highly COX-2–specific, did elicit questions; the human pharmacokinetics presentation also elicited a question of how to relate the recommended doses to the plasma levels and clearance.

The clinical program, presented by Steven Geis, M.D., encompassed fifty-one studies and over 13,000 participants. The five pivotal osteoarthritis trials and the two pivotal rheumatoid arthritis studies providing the data to support the sponsor's claim that celecoxib was comparable to naproxen, the most widely prescribed NSAID, were presented, followed by the results of the acute and short-term pain studies showing that celecoxib was more effective than a placebo.

Dr. Geis's presentation of Searle's human safety program began with the general safety studies, plus the particular analyses of adverse effects of celecoxib involving the kidney and nervous system, because COX-2 had been found in those tissues and blood platelets and because NSAIDs can have effects on the ability of platelets to function in blood clotting. This then led into the presentation of the studies of the effects of celecoxib on the gastrointestinal tract. The presentation focused on the five studies in which endoscopy was used to detect ulcers and the entire safety analyses for clinically serious symptoms of gastrointestinal side effects. The argument was set up by reference to a Searle-sponsored study of the impact of misoprostol, a synthetic prostaglandin, on the occurrence of serious upper gastrointestinal symptoms in elderly patients taking NSAIDs that was published in 1995 (Silverstein et al. 1995). This study of more than 8,000 subjects, half of them receiving misoprostol, not only demonstrated that those taking misoprostol had a 40 percent reduction in serious upper gastrointestinal events such as bleeding, ulcers, or perforation, but also that there was a 57 percent reduction in endoscopy-confirmed ulcers. This study, called the MUCOSA study, is later discussed as confirming the intuitive idea of the ability of endoscopy to predict serious GI events, thus supporting the use of endoscopy ulcers as a surrogate endpoint. Because serious GI adverse effects occur in only a small percentage of patients taking NSAIDs, acceptance of a surrogate that occurs more frequently, such as ulcers detected by endoscopy, might allow smaller trials to be used to demonstrate gastrointestinal safety.

The celecoxib GI safety studies, including five endoscopic studies, were presented, showing that the incidence of ulcers with celecoxib was similar to a placebo and statistically lower than the NSAID comparator, generally naproxen. This supported the contention that the NSAID GI label should be changed for celecoxib. The incidence of serious GI adverse events with celecoxib was significantly less than with the NSAID comparators but also higher than with the placebo. The problem with this analysis is that, though the safety analysis involved over 5,000 subjects and 5,000 patient years, for each study the number of serious GI adverse events was too small for any differences to reach statistical significance. The studies were not designed or powered (large enough) for that, which is why the acceptance of the surrogate was critical for Searle's goal of a significant modification of the NSAID GI label template.

The FDA presentation opened with a discussion by James Witter, M.D., the senior medical reviewer, of the real-world consequences to patients of the risks of NSAIDs, with many drugs often being taken by those taking NSAIDs, many other health conditions present, and the availability of NSAIDs, including aspirin, over the counter. He also differentiated the public hype about the COX-2 inhibitors as a Holy Grail for those with arthritis by noting that arthritis studies showed efficacy no better than for naproxen or diclofenac, meaning that the patients were still experiencing pain; it was just less pain. He noted that, though the suggestion was made to distinguish celecoxib from NSAIDs, the indication sought for both types of arthritis was the narrow indication of reduction in "signs and symptoms," not the more clinically challenging indications of prevention of structural damage, prevention of disability, induction of remission, or a complete or major clinical response. Dr. Witter also pointed to evidence from the sponsor's data from the open label portion of the arthritis studies of dose creep—patients taking more than the recommended dose of 200 mg twice a day to seek greater relief.

The FDA's presentation of the pain studies concluded that the results were at this time insufficient to support approval for the acute pain indication. The FDA reviewer concluded that the pattern and frequency of increase in blood pressure and fluid retention and more serious adverse events involving the kidneys in subjects receiving celecoxib were greater than for a placebo and typical of an NSAID.

The FDA reviewer's presentation on the effect of celecoxib on gastrointestinal damage was thorough and to the point: (1) The ulcer rate as detected by endoscopy was statistically lower than in the active comparator in several studies, though clearly higher than that seen with a placebo; and (2) celecoxib "may produce fewer clinically significant upper GI events than currently marketed NSAIDS" but (3) the studies submitted do not test that claim.

During the portion of the hearing given to comments from the public, four speakers representing different organizations had an opportunity to address their concerns to the committee. The first, Robert Palmer, M.D., of SmithKline Beecham, reiterated his argument presented at the March meeting that enzyme selectivity was insufficient evidence of safety, that the distribution of the two forms of COX enzyme was not as simple as argued by the Searle presentations, and that several lines of evidence suggested that inhibition of COX-1 was not responsible for the gastrointestinal damage of NSAIDs. Bill Soller, Ph.D., of the Nonprescription Drugs Manufacturing Association urged the committee to make sure the labeling for celecoxib accurately reflected its utility and safety and did not allow "unscientific" language disparaging the many useful and safe OTC pain relievers so that the media and patients are not misled. Tim Bryant, J.D., M.D., of the Aspirin Foundation of America, a nonprofit group supported by several large international companies marketing aspirin, urged the committee to consider the benefits and features of celecoxib on its own and not with regard to its superiority to other agents. He was particularly concerned, and understandably so, that the labeling not support marketing language such as super aspirin, suggesting that the "benefit to risk ratio" of available products, including aspirin, was unacceptable (Aspirin Foundation 2004). Finally, Sidney Wolfe, M.D., director of Public Citizens Health Research Group, also raised concerns that there was insufficient evidence to translate enzyme selectivity to safety, describing how serious GI side effects of another COX-2–selective drug, meloxicam, approved in the United Kingdom in 1996 and initially promoted as showing no damage to the stomach lining at twenty-eight days, later required the British regulatory authorities to distribute warnings to physicians that there was no evidence that there was a lower risk of the "severest" gastrointestinal adverse events with meloxicam than with use of other NSAIDs given at effective doses. Doctor Wolfe urged that long-term studies comparing the clinical safety of a new NSAID to ibuprofen, which he claimed is the "least dangerous" of the group, would be needed before removing the NSAID class GI warning. Ungraciously, he also noted that SmithKline Beecham (the company that had suggested that COX-2 selectivity was insufficient to establish efficacy and improved safety) had twice been found in violation of the FDC Act for claiming that its NSAID drug Relafen was safer because it was a preferential COX-2 inhibitor (CDER 1998d).

The afternoon session, given to the discussion and voting by the committee, went smoothly at the outset, providing unanimous approval for the OA and RA indications, with only a few questions asked and answered. The pain indication question sparked a lengthy discussion, with responses by Dr. Needleman and by Roland Moskowitz, M.D., a rheumatologist and consultant to Searle, defending the short-term OA study as a model for pain and the power of the

results with celecoxib in that model as supportive of the pain indication. The problem turned on the demonstration of efficacy in acute
pain after dental surgery and OA but not in the post-orthopedic surgery study. Committee members and FDA staff saw the signs and
symptoms of OA as consequences of inflammation and not an accepted model for pain. Dr. Needleman tried to get the FDA to allow
specific language in the pain indication, dental pain and OA pain, but
the reviewer held firm that the precedent is that the FDA requires two
studies to support approval for pain, and the OA study is not con-
tributory. The decision of the committee was not to decide; by consensus, they held that the standard for the pain indication, as stated in
the FDA's guidance document, had not been met and that more discussion and probably more data would be needed for approval for the
pain indication (CDER 1992).

The third question did not challenge but built on the view from
earlier Arthritis Advisory Committee meetings that endoscopic
studies are surrogates for clinically meaningful endpoints in the assessment of GI toxicity. The FDA asked, given that celecoxib had
demonstrated statistical superiority in the endoscopy studies to only
two of the three NSAIDs studied, what comparisons, if any, should
be allowed in the labeling between celecoxib and these NSAIDs, and
could the data be extrapolated to make comparisons between celecoxib and all other NSAIDs? This focused the discussion on whether
the results with endoscopy were the same with all NSAIDs. The
Searle team went through published data and their experience to argue that all NSAID's were the same, causing ulcers detected by endoscopy in about 20 percent of patients, and that the three drugs they
chose as comparators—ibuprofen, diclofenac, and naproxen—were a
reasonable group. They noted that this opinion had been shared by an
FDA representative during discussion of the trials. But one committee
member, Dr. Denis McCarthy, the nonvoting guest expert gastroenterologist, brushing aside the niceties of the FDA's measured approach
to the central issue, responded with a thorough presentation on why
and how endoscopy was not a sufficient surrogate. Drs. Geis, Silverstein, and Needleman defended their data manfully, providing details
on both endoscopy and serious GI adverse events with NSAIDs in the
published literature, the MUCOSA study, and the celecoxib trials, but

the game was on. The arguments meandered, but the question was called when a member of the committee, Dr. Daniel Lovell, a pediatric rheumatologist with significant clinical research experience, suggested that the large amount of endoscopic data should be included in the celecoxib label. And despite Dr. Needleman's efforts to return to the logic of the use of the endoscopic surrogate, the "you can't have a bleed without an ulcer" defense, in the end the decision was to include the data but otherwise retain the NSAID GI template, while noting that there was limited experience with the new drug. There was more parsing of the label and discussion of whether this was a new class of drug. Dr. Needleman pushed to get some reading on what would be required to remove the NSAID GI warning, and the answer was a "big study" with equivalence to a placebo for one or more clinically important GI events. Although the counterargument was made, based on the MUCOSA study, that such a trial would be prohibitively large and expensive, the chairman closed the discussion by voicing a consensus that the NSAID GI template remain but with modification to include the endoscopy results and removal of any language that does not pertain to celecoxib.

At that point, it was clear that Dr. Needleman was getting tired. After a weakly humorous exchange with the chairman about more slides and a suggestion from the FDA that further discussion focus on questions 5 and 11, leaving the FDA to work with the company on questions 6 through 10, Needleman said, "We surrender."

Question 5 dealt with the issue of the use of aspirin along with celecoxib, an issue of some financial importance with the growing use of low-dose aspirin to reduce the risk of strokes and heart attacks. Despite the data showing that celecoxib had no effect on platelet function, the committee chose to maintain the language used in the approved NSAIDs that low-dose aspirin can be used concurrently but that the combination may increase the risk of ulcers or other complications. In response to the final question, requesting input from the committee on what Phase 4, postapproval studies should be recommended, the committee recommended pediatric studies as well as the GI toxicity studies previously discussed. There are no publications indicating that Searle performed any pediatric studies, and the current label states that safety and efficacy in individuals under age 18 have not been evaluated (Pfizer/Pharmacia 2005).

So Searle and Phillip Needleman did not achieve all that they sought at this regulatory debut for celecoxib, but the game was not over. The committee and the FDA had provided guidance on what would be needed to gain approval for the pain indication and to get significant alteration in the NSAID GI template on the celecoxib label. These sequels will be picked up in Chapter 17. The draft label for celecoxib, with its new brand name Celebrex, was posted on January 1, 1999, and the drug was approved for sale on December 31, 1998.

For both drugs and both companies, now the challenge is to build on the years of research in the laboratory and in the clinic, the investment in manufacturing facilities and processes, and the grinding work of assembling, checking, and rechecking the thousands of pages of the NDA and launch the drugs for sale. This challenge, though it punctuates with a sense of triumph the many years of work, is only a beginning. Now the things that the business guys have been saying at all those project team meetings are tested.

16

The Launch: Marketing the Drug

All of the work to gain approval for a new drug is squandered if the drug does not sell. Although the origins of marketing new drugs may date to the snake oil practices of the patent medicine days, the crisis-punctuated evolution of the regulation of the industry and the maturation of the technology of drug discovery have led to a sophisticated effort to create customers and meet their needs, or at least their expectations. The purpose of marketing is "to reach customers and compel them to purchase, use and repurchase your product," a goal best achieved by attention to the four Ps: product, place, price, and promotion (Hiam 1997; Smith 2002). We have the product(s), the drugs that have achieved approval for sale by the FDA, and now will consider the place, price, and promotion as the drugs are launched, first made available for sale. Optimizing product placement is the challenge for those responsible for marketing the drugs.

As we have seen, the influence of marketing issues is felt early in the development of a new drug. The decision to develop a new drug, or even to invest in research required to discover a new drug for a particular indication, generally involves marketing research to define an unmet medical need, how many people have the medical condition the drug may address, the way the drug may be administered (pill, injection, etc.), and how it will be packaged and priced acceptably to both physicians and patients. There are many sources for such statistics, including the U.S. Centers for Disease Control and the World Health Organization, plus a range of proprietary fee-based sources of this information. Drilling into their data allows a company to understand how many people have the medical condition that may be treated by the new drug and where they are located geographically and structurally within the health care system (i.e., in the hospital or not, with private insurance or receiving state or federally subsidized health insurance, in an HMO, etc.). These numbers define the market in terms of the influences and constraints on prescription drug buying decisions. Later, these issues will also influence how the drug product is formulated for use (pill, injection, cream, nasal spray, etc.) and how it is packaged and priced.

The market research effort is not only quantitative but also qualitative. Who are the decision makers for treatments for this indication? What are the current options for treatment? What are their favorable attributes and their failings? If a drug is effective in treating osteoarthritis but must be given in a physician's office or clinic once a week, how willing or able are patients to make the trip each week and the physician to organize his or her practice to accommodate this? Will private or public insurance reimburse for the drug? Will the administration be reimbursed? Will the drug be put on the formularies, the lists of acceptable drugs, by hospitals and HMOs? How much better than current treatments in terms of efficacy and safety would the new drug have to be to make the cost of the drug and its administration cost effective for all those paying for the care of patients with the targeted indication? Would the drug save money for the health care system? If not, would the improved efficacy and safety override the added expense?

Clearly, the indication for the drug, the condition it is intended to treat, is a factor in this calculation. A drug that significantly extends the life of patients with cancer compared with the currently available therapies would likely be seen as sufficiently valuable to offset many of the issues of cost or inconvenience and create demand. But that is not an absolute. How long would the life extension have to be to justify the financial cost to the health care providers and the patients? What about the nonfinancial costs to the patient? If administration of the drug is painful and associated with disabling side effects or significantly decreased quality of life, how is the benefit weighed? These are some of the most challenging issues confronting patients, their health care providers, and the financial gatekeepers for new treatments. Although the issues surrounding a new drug are not always so perplexing, these difficult choices inform both the discussions and the decisions of regulators and their medical advisors to a greater or lesser extent when a new drug dossier is submitted for review. They also create the science-driven yet risky context for those charged with marketing and actually selling a new drug.

In a general sense, marketing, or product placement, is the work of overcoming the barriers that separate producers from consumers. The barriers include (Smith 2002):

1. Space—The product must reach the consumer.
2. Time—The customer may need the product at a time different from when it was produced.
3. Information—Producers need to know who the consumers are, what they need, and when and where they need it, and consumers need to know what is available, who has it, where it is, and what it costs.
4. Values—Producers are concerned with costs and prices, and customers value products for their utility and their reasonable cost.
5. Ownership—Producers own a product they do not want to consume, and consumers want the product owned by the producer.
6. Quantity—Producers benefit by producing and selling large quantities, and consumers buy and consume in small quantities.
7. Assortment—Producers may produce a limited array of products, and consumers need or want a large array.

To address the issue of place, marketing researchers provide the intelligence needed to plan and carry out the timely production and shipment of the product by the manufacturing and distribution arms of the company. Because of the barriers of time, quantity, and assortment, drugs are rarely distributed directly from the manufacturer to the consumer. The physical and financial path from the producer to the patient begins with a highly developed and regulated distribution industry. The Healthcare Distribution Management Association reports 85 U.S. drug distributor companies with over 200 distribution centers (i.e., warehouses), a fragmentation belied by the data on consolidation (HDMA 2003). In 1998, five large national wholesalers accounted for 90 percent of the wholesale drug market (FDA 2001a). By 2004, two mergers had reduced the "big five" to three companies that dominate the distribution of drugs to retail chain pharmacies, including mass merchandisers and food stores, independent pharmacies, and health care institutions (hospitals, clinics, and nursing homes). Smaller national and regional wholesalers pick up the balance of the business. Some wholesalers specialize in drugs of certain types: drugs that must be injected, biotechnology drugs, or drugs for certain conditions.

Suppliers have adjusted to meet the needs of the industry as prices for drugs have risen and as rules for dispensing a drug to a patient at or near the expiration date have tightened. The expiration date is established by stability tests done by the manufacturer on the drug substance (the active ingredient) and the drug product (CFR 2005f). When a manufacturer raises the price for a drug, a distributor with a drug purchased at the earlier lower price may be able to undercut the competition and still make a profit. This maneuver requires that the intermediate distributor carefully assess both the potential for the price to increase and also the cost of purchasing and holding inventory with expiration date clocks ticking. Conversely, sometimes a manufacturer will temporarily drop the price of a drug to increase sales, and suppliers that can purchase and distribute such drugs will benefit, again gambling on the capacity of their customers, the wholesalers, and their customers, the chain pharmacy warehouses and the pharmacies, to dispense the drug before the expiration date.

To make their drug purchases more efficient, hospitals and other institutions have developed large buying groups that negotiate a contract with a manufacturer for specific drugs at a discounted price for members of the group and to be filled by a prime vendor, a wholesaler. The wholesalers purchase the drugs at a nondiscounted price, warehouse them, and supply the requested drug under the contract at the discount price plus a service charge. The difference between what the wholesaler paid and the discounted price paid by the hospital is charged back to the manufacturer. This system provides quick service for the hospitals while freeing both the hospital and manufacturer from having to warehouse the drugs and freeing the manufacturer from having to maintain accounts with many hospitals (Sieker 2002). The power of the buying groups and other organizations that provide drugs to patients has contributed to changes in the rules about expiration dates. Hospital buying groups have moved to requiring twelve months of unexpired dating in their bid proposals, and the World Health Organization, the Red Cross, and other international charities have set a twelve-month minimum time to expiration for donated drugs (Sieker 2002; WHO 1999). Also, state pharmacy boards have moved from requiring that a dispensed drug need only be unexpired at the time of dispensation to requiring that the drug not expire before the patient completes the prescription (Sieker 2002). So-called "discrepancy distribution specialists" have arisen that match a customer's ability to dispense with the stock of distributors (Sieker 2002).

The Prescription Drug Marketing Act of 1988 (PDMA) was enacted to assure that prescription drugs are not counterfeited, misbranded, adulterated, or expired in this complex and shifting distribution system. Implementation by the FDA of amendments to PDMA passed in 2000 and intended to more rigorously document the "chain of custody" of prescription drug shipments as they move from the initial distributor to secondary regional and local distributors has been delayed (FDA 2001a).

Who finally provides drugs to patients? Most drugs are dispensed at chain drug stores or independent pharmacies. Hospitals account for 10 percent, mass merchants (e.g., Wal-Mart) and supermarkets about 12 percent, and mail order and online pharmacies less than 5 percent (FDA 2001a; NACDS 2003). In every instance, a state-licensed pharmacist fills the prescription, or drug order, of a physician. Pharmacists must meet the educational and continuing education requirements of the state board of pharmacy (NABP 2004). The state boards also define how drugs must be stored, the criteria a prescription must meet before it may be filled, what must be on the prescription label, what advice and information a pharmacist may or must provide to the patients, and what records must be kept. The state boards also set the rules for how a pharmacy will be equipped, run, and staffed (Sieker 2002). These state rules, as well as the federal laws governing the manufacturing and distribution of drugs, are considered in the decisions made about how to formulate, package, and distribute a prescription drug.

As an element of promotion, physicians provide free samples of drugs to patients. Pharmacies and pharmacists may not dispense or even possess drug samples.

The price of a prescription drug is also an element in product placement, locating the drug in the landscape of drugs that are available for a particular condition. Recall that a physician may legally prescribe an approved drug for any indication, any condition, though insurance or other third-party reimbursement agencies may or may not pay for such "off-label use." Manufacturers price prescription drugs by their assessment of the value to the patient after considering a number of issues, including the prices and properties of the competition, the features, both positive and negative, of their product that will lead a patient and physician to choose the drug, the nature of the indication on the label, the potential for expanded labeling with postapproval studies, the company's needs for revenue and market position, the company's ability to support the product with promotional activities and postapproval trials to expand the labeled indications, the current and anticipated issues of insurance reimbursement, and the public policy environment.

The price the company sets, the sticker price, is not what all, or perhaps most, patients or their insurance or health care providers pay for a drug. There are essentially five ways a prescription reaches a consumer and is paid for and, in each of these channels, business practices and federal law influence the price the final purchaser or consumer pays for a brand name prescription drug. The five customers are:

- cash customers, with no third-party payer;
- insurers and pharmacy benefits managers;
- HMOs buying directly from the manufacturer;
- Medicaid; and
- the federal government (e.g., VA hospitals).

Each channel has a particular set of processes and pressures that alter the list or average wholesale price set by the manufacturer. Pharmacy benefits managers, companies that manage the prescription coverage of health insurance plans, arose in the 1970s and 1980s as more and more private insurance added prescription drug coverage, and they have significant purchasing power. They process claims, provide mail order pharmacy services, negotiate for rebates with manufacturers, manage formularies and pharmacy networks, perform drug utilization reviews, work to encourage generic drug substitution, and most recently provide specific disease management programs for the covered individuals (ERG 2001). Health maintenance organizations (HMOs) that buy directly from manufacturers also have purchasing power that may translate into price reductions. The federal and state governments, through a variety of programs such as VA hospitals, Medicare, and Medicaid, are significant providers of health care and thus purchasers of drugs. Medicaid, administered by each state and paid for through the federal Center for Medicare and Medicaid Services (formerly HCFA), spent nearly $9 billion in 1997 on drugs, 17 percent of the national total (Brown et al. 1997). In 1990, the Omnibus Budget Reconciliation Act required that manufacturers wishing to have their drugs provided by Medicaid enter into contracts with states, collect and report the price of each drug sold, and agree to rebate to the Center for Medicare and Medicaid Services an amount determined by the average manufacturer price (AMP), the average price paid by all purchasers, and the best price to any. As of 1996, the rebate was the larger of 15.1 percent of the AMP per unit or the difference between the AMP and the best price per unit, adjusted by the consumer price index on the launch date and the current quarter's AMP (CMS 2003). The effect of the negotiated prices, negotiated rebates, and legally mandated rebates is that the amount paid for a prescription drug by the final purchaser ranges from a little over the list price (cash customers) to less than half the list price (federal hospitals) (ERG 2001).

Given the complexity of the sales and distribution channels and the many competitive and legal influences on what a company may recover for the sale of its drugs, there is no magic formula to pricing, just the experience and judgment of the professionals using the extensive research available to them.

"Megabrands," products with potentially large markets, treating medically homogeneous patients chronically, will have included head-to-head comparisons with the dominant product in their preapproval trials, particularly when there is steep competition for market share, perhaps one or a small number of existing products (Kessler et al. 1994). The launch strategy will be based on the benefits, especially for chronic use, of the new drug. Because the market may be large, the price per unit may be low (Kevin O'Neill, personal communication 2004).

Specialty drugs, addressing smaller markets, perhaps with subsets of patients, and generally high unmet needs and no cures, will have had a very focused clinical development to achieve fast entry into the market, particularly if there are no or few approved competitors or if there are other potential competitors in development (O'Neill, personal communication 2004). Advances in biomedical sciences, providing new understanding of the disease and/or opening up new opportunities for therapies, often result in a clutch of products reaching development at nearly the same time. The prices for such specialty drugs often are high, reflecting not only the cost of development and production but also the risk of moving into an unknown and small market. The absence of available alternatives will also support higher prices, a sad but true aspect of a for-profit drug industry.

Promotion addresses the mental landscape of product placement, providing the information to the drivers of drug choice that will lead them to select one drug or another. The formal customers of prescription pharmaceuticals, at least the legal gatekeepers, are physicians, but patients, the literal consumers of drugs, are becoming significant drivers of demand for prescription drugs. Pharmaceutical promotional activity includes advertising in medical journals, sales visits to physicians' offices and hospitals, direct-to-consumer marketing, and the item of greatest expense, the retail costs of drug samples provided to doctors for their patients. From 1996 to 2002, promotional spending grew from $9.2 billion to $21.5 billion, with the retail value of samples doubling to account for over half of promotional spending. Spending on promotion by pharmaceutical companies is significant and growing not only in absolute terms but also in relation to research and development (R&D) and as a percentage of sales revenue. Between 1987 and 1990, marketing expenses consumed between 17 and 26 percent of sales (OTA 1993). In 2001, R&D spending industry-wide was estimated at $30 billion and promotional spending $19 billion. Spending on R&D increased 30 percent between 1999 and 2001 and promotional spending 35 percent in the same interval (PhRMA 2004). Spending on promotion by ten of the largest pharmaceutical companies in 2002 ranged from $0.75 to over $3 billion. Pfizer spent $3 billion on promotion, an increase of 27 percent from the prior year (Lawrence 2003).

Promotion to the gatekeepers, the physicians, consumes the largest portion of the marketing budget. However, expenditures on advertisements in medical journals, visits to physicians' offices and hospitals by salespersons, and booths at medical conferences consume less than half of the marketing budget for a prescription drug. The single largest element of the promotional spending for many drugs and the industry as a whole is the provision of free samples to physicians, expensed at the retail price of the drugs, which totaled over $6 billion in 1998 and grew to $11 billion in 2001 (Ma 2003; PhRMA 2004). Promotion in almost any form—information, samples, or coupons—can be thought of as either push marketing, with the initiative entirely in the hands of the promoter, or pull marketing, with the initiative entirely in the hands of the intended purchaser. The practice of "sampling" had been essentially push, with the sales representatives, sometimes called detail men and women, dropping off or mailing samples to a physician so that the physician would have the opportunity to easily and efficiently do a quick test to see whether the new drug would help his or her patients as advertised. The sales reps receive extensive and ongoing training on the business and regulatory aspects of pharmaceutical marketing, on the products they are discussing, or detailing, with the physicians and the basics of the diseases the drugs address (Ferris, personal communication 2004). They provide information and samples.

In 2000, the FDA, in response to reports of a black market in diverted samples, developed tighter rules on the distribution of drug samples (Romanski 2003). These rules require a written request for specific prescription drug samples from the physician before the delivery by hand or by mail of drug samples (CFR 2005f). The company may provide preprinted forms for the request, but the physician must sign and date a request and specify not only the drug but also the dosages and the amounts. The physician also must sign a receipt for the samples upon delivery. Because the request and receipts provide a paper trail for drugs that would otherwise not be accounted for by the usual commercial distribution channels to pharmacies and hospitals, this was a reasonable exercise by the FDA for the control of access to prescription drugs, and it might have been expected to put a bit of a governor on the "push" marketing by samples. Yet the dollars spent on sampling grew by over 50 percent from 1999 to 2001 (PhRMA 2004). The average retail price for prescription drugs grew, though less substantially, during the same period; brand name prescriptions on average grew from $53.51 in 1998 to $65.29 in 2000 (Kreling 2001). This suggests, despite the new regulatory requirements, increases in the volume of samples provided drove the growth in expenditures for this form of promotion.

Another aspect of promotion by pharmaceutical companies to physicians is the giving of gifts, from small items such as notepads or pens to large gifts such as free trips to vacation resorts for educational or non-educational events and dinners at high-priced restaurants, followed by presentations on a company's drug and cash fees for providing minimal consulting services. Sales representatives have often provided a free lunch for hospital educational events, textbooks and black bags for medical students, and scholarships for students and residents to attend medical conferences. The frequency and pervasiveness of such gifts, which were perceived by some as inducements for prescribing a particular drug or even as a kickback for doing so, became a topic of some notoriety, and even the literature on the topic became a cottage industry in the 1990s (Wazana 2000). In the early 1990s, there were Senate hearings on the issue, and several professional medical organizations and even the Pharmaceutical Manufacturing and Research Organization formulated and published ethical guidelines on gifts from pharmaceutical companies (ACP 1990; AMA 1991; Katz and Caplan 2003; Randall 1991; Reynolds 2002). Some organizations, such as PhRMA, have made distinctions between large and small gifts or between gifts that may be perceived as potentially benefiting patients or not. For example, a notepad emblazoned with the logo for a drug is okay, but a tote bag similarly decorated is not. A patient education poster on the biology of high blood pressure is acceptable, even one gaudily decorated with the logo of a treatment for the condition. Bioethicists have suggested that the value of the gift is an untenable basis for distinguishing between or among gifts intended to influence the prescribing actions of the physicians (Katz and Caplan 2003). That is precisely what the companies intend, and although it is not wrong to wish to influence those in the decision-making role in the purchase of one's product, the premise of the relationship between physicians and society, between any professional and society, is that decisions will be based on the best professional judgment of prescribers and not in response to gifts, favors, or inducements.

The available research supports the contention that certain interactions between the pharmaceutical industry and physicians influence physician behavior; the provision of samples increases the likelihood of rapid prescription for a new drug, and payment of honoraria or funding of research increases the likelihood of a request that a new drug be added to a hospital's formulary, the list of drugs routinely available from the hospital pharmacy (ASHP 1983, reviewed in Wazana 2000). A study published in 1992 documented that the use of two drugs for hospitalized patients with lung disorders increased severalfold after hospital physicians had attended symposia in vacation resorts about the drugs (Orlowski and Wateska 1992). Despite such data, physicians when asked will generally express doubt that gifts, trips, or samples change their prescribing patterns (Goodman 2001). It would require a thorough study of patient outcomes to know whether these changes in prescribing or advising activities benefited or harmed patients, and such data are not available (Wazana 2000).

In 1993, a pharmaceutical marketing firm tested the relative power of different forms of physician-directed promotion methods not on sales but on the ability of a physician to understand the next promotional message. Dinner meetings with presentations were five times more effective than a journal advertisement, and preprinted prescription pads or notepads were one-third as effective as a journal advertisement. Visits by sales representatives were not as effective as a dinner but still were found to be over three times as effective as a journal advertisement. So, even absent data connecting a particular form of promotion to sales, these kinds of results support intensive and carefully crafted marketing strategies to physicians (Smith 2002).

The impact of marketing to physicians goes beyond the prescribing practices of individual physicians, particularly for insurance companies, health care systems such as managed care organizations, health maintenance organizations, and hospitals. These institutions and organizations establish a formulary, the list of pharmaceuticals that reflect the current judgment of their medical staff, which undergoes constant revisions, based on the requests of prescribers and economic as well as safety issues that arise. An institutional pharmacy and therapeutics committee, composed of physicians, pharmacists, nurses, and administrators, reviews and approves the recommendations based on assessments of efficacy, safety, and economics, a process that may include requests to manufacturers to submit new bids to identify the lowest-cost agent. Pharmaceutical marketing efforts therefore include hospital pharmacies, providing information useful for the evaluation of their products.

Consumers are also increasingly targeted by promotional messages for prescription drugs, a fact apparent to anyone who has turned on commercial TV in the last few years. The parade of spots for prescription and over-the-counter drugs escorting the nightly news makes one wonder about the health and well-being of the viewing audience at 6:00 P.M. Pharmaceutical companies spent nearly $4 billion on direct-to-consumer (DTC) advertising in 2002, much of it in the form of radio and TV broadcast ads (Lawrence 2003). In 1994, the total spent on DTC promotion by pharmaceutical companies was $266 million, 13 percent of which was for TV ads (Kreling 2001). In 2000, $2.5 billion was spent on DTC promotion, about 2.5 percent of what was spent on mass marketing of nondrug consumer products (NIHCMF 2000). The growth of DTC promotions, an average of 44.9 percent each year from 1994 to 2000, is in part a response to increased patient autonomy and the increasingly easy access to health and health care information through the Internet and other channels (Kreling 2001). DTC marketing of prescription human and animal drugs is regulated and monitored by the FDA, specifically the Division of Drug Marketing, Advertising, and Communications (CFR 2005e). Print advertisements must include a summary of effectiveness, side effects, warnings, contraindications (who should not take the drug), and precautions that is a balanced and "true statement" as approved in the label. What constitutes failure in presenting a "true statement" is described in detail, and such failures are violations of the federal Food, Drug, and Cosmetic Act, section 502(n), and subject to substantial civil penalties (Federal Food, Drug, and Cosmetic Act, as amended by the FDA Modernization Act of 1997). The FDA has published a Guidance on Consumer-Directed Broadcast Advertisement Contributing that notes that radio and TV advertisements for prescription drugs as well as telephone marketing to consumers must provide the "major statement," the product's major risks in either the audio or audio and visual parts of the ad, and provide a brief summary of the approved label or make what is termed an "adequate provision" for the dissemination of the approved package label in the form of toll-free numbers or the URL for an Internet site (FDA 1999). In response to phone or electronic requests, the label may be provided by mail, or if the individual is concerned about privacy, the company must provide a toll-free number so that the label can be read to them. An acceptable alternative is a referral to sources of the label in print advertisements in magazines or other publications. The system of FDA oversight of DTC advertising of prescription drugs was assessed in 2002 in a report by the General Accounting Office of the U.S. Congress to a bipartisan group of three senators and two representatives. The GAO found that though the regulatory letters sent by the FDA to manufacturers who disseminated misleading advertisements, which accounted for about 5 percent of those advertisements they reviewed, were effective in halting the ads, the companies do not always submit advertisements for review. The GAO also found that a change in policy at the Department of Health and Human Services, which required a review of regulatory letters by the FDA's Office of the Chief Counsel, had so slowed the issuance of letters that many were sent after the broadcast life of the advertisement (GAO 2002b).

Other federal agencies have begun to pay attention to DTC promotion of drugs, including the Federal Trade Commission and the Office of the Inspector General (OIG) of the Department of Health and Human Services (Kalb et al. 2003). The FTC, acting on its mandate to prevent deceptive and unfair marketing, has begun to move against online pharmacies. The FTC has also begun to investigate whether material sent to consumers as part of patient education and patient compliance programs of large pharmacy chains contained false or misleading statements about the safety or efficacy of drugs (Kalb et al. 2003). The OIG has raised concerns that payments from manufacturers to pharmacy benefits managers to assist in DTC marketing may violate antikickback laws (Kalb et al. 2003). The antikickback law prohibits direct payments for switching of products in federally reimbursed programs, but the OIG now suggests that payments for DTC marketing may constitute indirect kickbacks (Kalb et al. 2003).

Results are inconsistent on whether DTC advertisements influence physicians' prescribing behavior. One small study found that 80 percent of people surveyed had seen an ad for a prescription drug and 25 percent discussed the ad with their physician, but only 6 percent received a prescription for the advertised drug (Rosenthal et al. 2002). Surveys reported at an FDA meeting in 2003 indicated that patients' requests for drugs they had seen advertised did have an impact on physicians' prescription choices, particularly physicians in training (FDA 2003e). There are no data to suggest that the influence damages or benefits patients or that a physician prescribed a drug against his or her better judgment. But the economic data and patient surveys do suggest that the investment in DTC advertising pays off and that patients feel that they benefit from the ads (FDA 2003e). The fifty prescription drugs most advertised to consumers accounted for nearly half ($9.9 billion) of the increase in prescription drug spending from 1999 to 2000; increases in the sales of the other prescription drugs (about 9,850) accounted for 52 percent of the increase (NIHCMF 2000).

The challenge to the marketing group in the period leading up to the assembly, submission, and review of the NDA, when the company and the team have a level of confidence that they have an approvable drug, is to put together a marketing plan with a budget that will provide the greatest return on the investment in promotion. Many of these activities and materials are, in their final details, contingent on the approved labels negotiated with the FDA after Advisory Committee approval. The packaging must be developed using stability studies and the requirements defined in the NDA. The packaging is tested with focus groups and the packaging plan is used to develop and carry out the production of the commercial product package, shrinkwrap and all. Representatives from the major channels for distribution are contacted and the market forecast developed. Advertisements and promotional materials for physicians and public media are developed and approved. The detailing plan must be developed and training materials developed and tested, even before approval. The training materials may need revisions based on the final label the drug receives from the FDA. Sales representatives must be trained. Plans for conventions are developed on the basis of scheduled professional conferences, and the exhibit materials that will be used in conventions are developed. Promotional slides, tapes, and videos are developed with contingencies for revisions required by the approved label (Smith 2002). All of these activities are coordinated by the marketing managers with an eye toward being prepared for the complex process of actually launching the drug for sale but with the flexibility to deal with any last-minute changes required by their managers or regulators in label, package, and promotional copy.

MARKETING OF COPAXONE

To distribute Copaxone, Teva in 1995 formed Teva Marion Partners with Marion Merrill Dow (now Aventis) (*The Jerusalem Post* 1995). The arrangement was somewhat complex. Teva manufactured the drug and sold it to Marion Merrill Dow, who contracted the marketing and distribution to Teva Marion Partners (Institutional Investor 1998). Teva eventually took over the marketing of Copaxone in the United States. The launch of Copaxone in the United States in April 1997 was preceded by projections of $100 million or more of sales per year, which would, at the price of $9,500 for a year's supply of the drug, translate into just over 10,000 patients. Copaxone was the third drug approved for the indication. Schering's Betaseron and Biogen's Avonex, both engineered forms of human interferons, immune system signal proteins, approved in 1993 and 1996, respectively, had slow launches and some problems, and competed with Copaxone. By fall 1996, 30,000 patients were using Betaseron, down from 60,000, and 15,000 were using Avonex. Both interferons cause flu-like symptoms for a period after dosing, and some patients and physicians felt that Avonex, with once a week dosing, was more tolerable. For both interferons, there was also a concern that antibodies formed to these engineered proteins would compromise efficacy (Rosenberg 1996). Biogen's marketing of Avonex to neurologists and the community of MS patients was so effective that by late 1996 it held more than half of the U.S. market (Johannes 1996). By 1998, there were 30,000 patients taking Avonex. Market penetration for all three approved MS treatments was low; only about one-third of the estimated number of people with the relapsing form of MS were taking one of the three treatments. Copaxone's first-year U.S. sales were less than hoped; 6,000 patients were taking Copaxone by March 1998, though half of the new prescriptions for one of the three approved MS drugs were for Copaxone (Solomon 1998). By 2001, 143,500 patients were receiving one of the three drugs. Copaxone, Avonex (53 percent), and Betaseron (21 percent) (Mehrotra et al. 2003).

In March 2002, Rebif, a new formulation and dose of the same interferon molecule as Avonex, was approved despite Avonex's orphan drug status because Serono, Rebif's sponsor, had shown Rebif's clinical superiority to Avonex in a head-to-head comparison (Haffner 2002). By mid-2003, 178,800 MS patients were being treated with one of the interferons or Copaxone, and estimates for the full year had Avonex's share at 40.9 percent, Rebif with over 10 percent, Betaseron holding at 21 percent, and Copaxone 31.9 percent (Mehrotra et al. 2003). Marketing for all four drugs was intense, and to add to its reach Serono established a marketing partnership with Pfizer, a company new to MS but with a formidable 8,000-member sales force (Vollmer 2002). Although the relatively small potential market for MS drugs would make mass-market DTC marketing unlikely to be cost effective, Schering, Biogen, Teva, and Serono/Pfizer, in addition to sending their sales representatives to the offices of the 5,000 neurologists who care for patients with MS, exerted DTC marketing by not only advertising in disease-specific magazines, such as Inside MS, the glossy publication of the National MS Society (NMSS), but also by sponsoring educational programs for people with MS. These programs are generally run in cooperation with local physicians and hospitals. Companies enroll leading neurologists, including the investigators leading their clinical trials, into speakers' bureaus and make them available for talks, nominally cosponsored/branded by hospitals and/or chapters of MS private agencies such as the NMSS. The drug company would pay the honoraria and travel expenses for the speaker as well as the costs for the space and any refreshments. The NMSS has worked very hard to assure that the talks it sponsors are balanced, eventually requiring that the chapter's advisory physicians vet the speaker and the content. This creates an understandable tension between drug companies' marketing arms and the NMSS, based on what each has to contribute and gain, the company providing an effective and credible speaker for resource-strapped chapters and the chapter providing access to its members, the company's potential consumers.

In addition to these efforts, Teva, as well as the other companies marketing the new treatments for MS, moved into customer relationship marketing. It was a natural extension of the initial training provided by nurses for these drugs that had to be injected daily, every other day, or weekly by the patient or a caregiver, such as a spouse. The company-trained nurse made an appointment and visited the patient's home, providing instruction and training on the appropriate way to mix and inject the drug safely and dispose of the syringe and needle. This established a relationship between the patient and the company that was educational and helpful rather than simply commercial. The companies expanded this to a form of customer relationship marketing (CRM), creating mail- and Internet-based educational programs for people with MS. Teva pioneered the use of the Internet to provide an online consumer disease community, launching MSWatch in July 1998. In addition to raising consumer awareness, driving conversions from one of the interferons to Copaxone, and improving patient compliance, the site allowed Teva to collect demographic and disease data to better understand their consumers. One challenge to these programs, which was tied to the patient's registering and sharing personal and medical information important to marketing, was the passage of the Health Insurance Portability and Accountability Act of 1996 (HIPAA), which contained language that required entities that provide and charge for health care services to develop systems to protect the confidentiality of personally identifiable material. Pharmaceutical CRM sites do not charge and thus are not strictly covered by HIPAA, but companies are becoming concerned that they must enhance efforts to build and maintain the trust of those to whom CRM is pitched and thus work to meet the HIPAA-recommended practices for confidentiality (Arnold 2003).

MSWatch, highly customized for people with MS, quickly enrolled over 25,000 users. By registering, consumers gained the use of health management tools, such as diaries to track symptoms and treatment. Nurse counselors are available, as are discussion groups in real time, forums, buddy lists, and white pages to share support and advice as well as feedback to the site. Users have organized offline meetings (Evenhaim 2001).

The other companies marketing drugs to people with MS have followed suit. Schering provided an unrestricted educational grant to fund Multiplesclerosis.com, Biogen established MS Active Source, and once Rebif was approved Serono/Pfizer established MSLifelines. These sites have medical advisory boards and a credible air of help, if not complete objectivity. As examples of CRM, these sites and support systems provide tools to help people with chronic diseases manage a difficult and less than curative therapy and encourage both healthy lifestyle choices and sticking with the therapy program. They are effective marketing and marketing research tools.

CELEBREX MARKETING

The Celebrex launch in mid-February 1999 was a carefully orchestrated triumph. In fact, MedAd News chose Celebrex as the 1999 brand of the year (Smith 2002). Searle had established a U.S. co-promotion agreement with Pfizer for celecoxib and a second-generation COX-2 inhibitor, which was quickly expanded to cover all markets except Japan (FT Asia Intelligence Wire 1998; Pharmaceutical Business News 1998). Searle, after extensive market research among consumers, physicians, and managed care organizations, and computer modeling of competitors, had set the average wholesale prices, $2.42 a pill for the 200-mg pill and $1.43 for the 100-mg pill, to be at or below the cost per month for the brand-name prescription NSAIDs, including their own

Daypro (Steyer 1999a; Rock 1999). Supplies were rolled out to pharmacies in late January to mid-February, after Searle/Pfizer sent 45,000 patients starter kits of ten bottles of a twenty-five–day supply of pills to physicians and pharmacies (Steyer 1999a). The mailing list for samples was aimed at those physicians and practices that prescribe NSAIDs, information the company had derived from their sales of Daypro (Steyer 1999a). Prior to a media blitz in medical journals, magazines, and newspapers, sales representatives were brought to San Francisco for four days of training. In its first five weeks on the market, 409,000 prescriptions were written for Celebrex, a launch pace bested only by Viagra (St. Louis Post Dispatch 1999). At fifteen weeks after launch, more than 3.5 million prescriptions had been written for Celebrex, and in the first year sales of Celebrex were $2.5 billion (Pham 1999; Smith 2002). Sales of Celebrex grew to over $3.1 billion, though total U.S. sales dropped 2 percent to $2.6 billion, in 2002 (Kumar and Zaugg 2003).

Vioxx, a second COX-2 inhibitor, developed by Merck, was approved four months later and by 2003 achieved large sales and a significant market share, though Celebrex remained the market leader (Anonymous 2003). Marketing and a head start may tell the story. The promotional expenditures for the first six months of each were $159 million for Celebrex and $111 million for Vioxx; Searle's first-year promotional spending for Celebrex was double Merck's for Vioxx, $310 million to $162 million. The mix of marketing was similar, with two-thirds on detailing. Direct-to-consumer advertising of Celebrex and Vioxx began in the last quarter of 1999, with both companies spending nearly 60 percent of their DTC budget on television (Liebman 2000).

Consolidation in the pharmaceutical industry has resulted in Pfizer having sole ownership of Celebrex. In late 1999, Pharmacia bought Monsanto. Eventually Monsanto's agricultural business was spun off as a separate company, and in 2003 Pfizer bought Pharmacia.

Two other COX-2 inhibitors were approved: Mobic from Boehringer Ingelheim in April 2000; and Bextra, Searle/Pfizer's second-generation drug, in November 2001. In their 2002 Annual Report, Pfizer reported that Celebrex was receiving 22 percent of arthritis prescriptions and Bextra 8 percent. Celebrex in 2003 was the ninth best-selling drug, and

COX-2 inhibitors accounted for $5.3 billion in U.S. sales (IMS 2004). This was big business indeed, despite little evidence of improved efficacy for any of these drugs when compared with older NSAIDs and only minor tinkering with the NSAID template on the label.

17

It's Not Over Till It's Over: Post-Approval Studies

Research and the collection of clinical data do not stop with the approval and launch of a drug. Regulations require a program of adverse reaction reporting to the FDA, and business considerations require the management of the life span of marketing exclusivity of a drug.

Because newly approved drugs have been tested in relatively small numbers of individuals, new safety issues may arise after the drug reaches the market. The FDA has established programs to continue to monitor the safety of approved drugs (CDER 2004a). In the course of postmarketing studies being run under an IND, adverse events are reported annually as required for any ongoing clinical trial, and SAEs in such studies must be reported within fifteen days of receipt of the information if it is reasonable to believe that the drug caused the adverse event. Holders of approved NDAs and manufacturers and distributors of approved drugs are responsible for postmarketing safety reporting of approved drugs. The manufacturer or distributor must report serious, unexpected adverse experiences associated with a drug to the FDA within fifteen days of receipt of the information (CFR 2002). Also, the NDA applicant of an approved drug must submit periodic reports of adverse events not reported under the fifteen-day rule quarterly for three years and annually thereafter (CFR 2002).

A physician, pharmacist, or other health care provider can report possible adverse effects of medical products to MedWatch, established by the FDA in 1993 (FDA 2004a). Reports may be submitted electronically, by phone, or by fax. After evaluation by FDA staff, the reports are compiled and entered into a database. As a result of reports, the FDA may issue safety alerts to the public, send letters to healthcare professionals (aptly called DHCP, dear health care professionals), require changes in the drug's label, or require the withdrawal of the drug from the market. MedWatch also compiles reports about serious adverse events or other problems with medical devices, dietary supplements, cosmetics, and infant formulas (Meadows 2003). MedWatch distributes the alerts and safety warnings through several professional organizations, an e-mail newsletter, and at their Web site (http://www.fda.gov/medwatch/safety.htm).

The problem with much of the postmarketing adverse event reporting system is that it relies on voluntary submissions of reports by physicians to the manufacturer or to MedWatch. Manufacturers must submit regular reports of serious adverse events associated with the use of a marketed drug, but they can only submit what they receive either by voluntary reports or if the event occurred during the conduct of a company-sponsored trial under an open IND. No serious analyses of the overall effectiveness of MedWatch in capturing serious adverse events have been made.

With the growth of contacts between manufacturers and consumers such as "company-sponsored patient support programs and disease management programs," companies are acquiring new types of safety information. For purposes of postmarketing safety reporting under federal regulations, the FDA has decided that reports of potential adverse experiences derived during such planned contacts and active solicitation of information from patients should be handled in the same way as safety information obtained from a postmarketing study in terms of the manufacturer's reporting obligation.

To attempt to harmonize rules for safety reporting on drugs internationally, The International Committee on Harmonization has developed guidelines for planning postmarketing safety monitoring, which they call pharmacovigilance (ICH 2005). The challenge is to have a system that has the frequency, manner, and duration of reporting for drugs that are approved at different times in different countries.

A search of MedWatch for alerts on Copaxone found only alerts for all the approved changes to the package inserts and the patient information sheets, including the additional labeling for preloaded syringes approved in the fall of 2001.

A MedWatch search for Celebrex from approval to 2001 yielded several hits, beginning with an alert in 1999 for people taking Celebrex and warfarin, a drug used to reduce the ability of the blood to clot in some people at risk for a stroke. In the Phase 1 studies, results from healthy subjects given both Celebrex and warfarin indicated that Celebrex did not increase warfarin's effect on blood coagulation. However, in postmarketing studies, some people taking warfarin experienced an increase in anticoagulation and risk of bleeding after starting Celebrex. The label was also changed in December 2000 to strengthen the language describing the effects on fetuses when pregnant rabbits received Celebrex at doses leading to two to six times higher exposure than the human dose of 200 mg twice a day. A precaution was added regarding the use of Celebrex in nursing women because a study in lactating rats and analysis of the milk of one woman indicated that Celebrex is excreted into the milk and may pose a risk to the nursing infant (Pfizer/Pharmacia 2002).

The subsequent alarms raised about Celebrex, Vioxx, and other NSAIDs, whether COX-2–specific or not, were in part an unintended consequence of continued efforts by the companies to expand the indication and thus the life span of their exclusivity. The business of managing the drug after approval is challenging and may be what distinguishes successful from unsuccessful drugs, franchises, and even companies. Although the commercial life of a new drug begins with its approval by a regulatory agency such as the FDA, the actual life span of a drug begins earlier with the filing of the patent application. Under FDA rules, the exclusive right to market a drug may be expanded if the drug is designated a new chemical entity or an orphan drug. As described in Chapter 5, unit sales and revenues drop to half the peak by six years after patent expiration and approach 0 at twenty years from patent expiration. Competition for the innovator drug arises from both the generic versions of the same drug and from new drugs for the same indication. Because the pharmaceutical industry is so profoundly R&D-driven, it is not surprising that the second type of competition, based on innovative alternatives to achieve the same biological effect, has a more significant impact on the slowing of the innovator's sales (Lichtenberg and Philipson 2002). These realities drive pharmaceutical companies not only to market aggressively but also to support more clinical research that will expand the label and possibly extend exclusivity. Companies also manage a drug's life cycle by research supporting patent applications that expand the intellectual property protection in time and scope.

MANAGEMENT OF COPAXONE'S LIFE CYCLE

Teva's efforts at managing the life cycle of Copaxone have relied in part on the complexity of the drug and the difficulty in synthesizing it to pharmaceutical standards. Over the years, Teva has sought and gained patent protection for improved methods for the synthesis of Copaxone and its production to more stringent standards of size. Generic glatiramer acetate is unlikely not only because the patent protection expires in 2014. There is another barrier to generic competition for Copaxone. One notable absence from the NDA was a method to measure the drug in the blood. Thus there are no human studies in the NDA or in the literature describing the time course of the entrance and clearance of Copaxone into the blood. The Hatch-Waxman Act allowed easier entry of generics by requiring the applicant to demonstrate only that they could manufacture the drug and drug product to the FDA's standards and that the body comparably handled the generic and innovator drugs. New human trials to demonstrate the clinical efficacy and safety were not required. But without a method to measure the drug in the blood, a comparison for how the generic and innovator are handled by the body is impossible. This does not mean that a potential competitor could not develop such a method, but it is not in Teva's interest to do so. The more general issue of how generic biotechnology drugs should be tested and approved is currently being debated and evaluated by the FDA.

A larger threat to Copaxone's market share is the expanding array of drugs for MS. Copaxone's share of the market, around 30 percent, has held steady for a few years, based on more or less comparable efficacy and more tolerable side effects than for the interferons (Betaseron, Avonex, and Rebif) (Maggos 2004). But in late 2004, Biogen-IDEC and Elan gained approval for and launched Tysabri, an entirely new drug for this disease. Tysabri is an antibody directed to proteins on the surface of white blood cells. Blockage of these proteins by the antibody prevents the cells from sticking to the inside wall of the blood vessels and moving from the blood vessel into the central nervous system. Tysabri was approved based on one-year efficacy results in two trials, one in which Tysabri or a placebo was given to patients with MS who were not taking any other MS drug and a second trial in which patients taking Avonex and experiencing relapses received, in addition to Avonex, either a placebo or Tysabri (Biogen-IDEC/Elan 2004). In the first trial, subjects taking Tysabri had a 66 percent reduction in relapses when compared with the placebo group. In the second trial, subjects adding Tysabri to Avonex experienced a 54 percent reduction in relapses compared with those patients receiving Avonex alone. Tysabri is administered in a doctor's office or clinic once a month through a vein in the arm in a procedure that takes an hour plus an hour's observation (Biogen-IDEC/Elan 2004). There were no flu-like symptoms as seen with the interferons, but allergic reactions and infections were seen with Tysabri in both studies. The potential impact of Tysabri on the market for MS drugs was not clear, though the marketing plans were fierce (Maggos 2004). It was expected to further fragment or expand the overall market. It was approved for patients with relapsing-remitting MS (~85 percent), either as the first treatment started or for those failing other treatments, and could be used alone or in combination with other MS drugs (Biogen-IDEC/Elan 2004). Biogen-IDEC/Elan committed to completion of the second year of both studies and an ongoing study with Copaxone. Preliminary reports indicated that, when used alone, Tysabri provided a 67 percent reduction in relapse rate and a 42 percent slowing of progression at two years. The retail price was set at $28,100 a year.

But in late February 2005, Biogen-IDEC/Elan withdrew Tysabri from the market and stopped all dosing with the drug in clinical trials, which were going on not just for MS but also for an inflammatory disease of the bowel and for rheumatoid arthritis. They withdrew the drug, after consultation with the FDA, when they learned that two patients who had been receiving Tysabri along with Avonex developed an often lethal neurologic disease called progressive multifocal leukoencephalopathy (PML). In PML, large regions of the brain lose myelin and cease to function as a result of the activation of a virus called the JC virus. The JC virus has infected more than 80 percent of us with no problems, but in individuals with profound immune system defects, the virus enters the brain and causes extensive damage; PML is most often seen in AIDS patients. Why these two individuals, who had received the combination for a bit more than two years, developed PML is not known. It is reasonable to speculate that the prevention of the entrance of immune system cells into the central nervous system by Tysabri contributed to the failure to control the growth of the JC virus. But why just these two, and why only in the group who were receiving both Avonex and Tysabri? The companies and the FDA are working to sort this out; it is unclear whether Tysabri will return to the market, even as a solo treatment. Soon after the Tysabri withdrawal, the FDA halted testing of a drug being developed by Glaxo for MS that targeted the same adhesion molecule as Tysabri, just as a precautionary move until it could determine whether Tysabri did contribute to the PML. One of the two Tysabri PML patients died, and the second was reported to be recovering. In early April 2005, the companies reported that another patient taking Tysabri had died of PML. This was a patient with the inflammatory bowel disease, whose death had originally been mistakenly attributed to a brain tumor.

Franchise management for large-market drugs can be more inventive and aggressive. Witness the case of Claritin and Clarinex. Clarinex is the active metabolite of Claritin, so as the Claritin patent and exclusivity were expiring, Pfizer submitted an application for the change for Claritin from prescription to over the counter and gained approval to market Clarinex, the metabolite, developed as a distinct drug (McClellan and Jarvis 2001).

MANAGEMENT OF THE CELEBREX LIFE CYCLE

Searle/Pharmacia/Pfizer expanded the Celebrex franchise with several regulatory filings, some moving through the FDA approval process with more ease than others. An NDA to allow marketing to reduce the number of polyps in patients with familial adenomatous polyposis (FAP) was submitted in June 1999 and approved in late December 1999. Because FAP, a rare (estimated to affect 1 in 10,000 people) inherited condition in which thousands of polyps develop on the inside of the lower portions of the digestive tract, generally results in colon cancer, FAP patients undergo surgical removal of much of their colon (NORD 2001). A growing body of evidence suggested that the COX-2 enzyme protein played a significant role in driving polyp formation in FAP patients, and a trial of Celebrex in 77 FAP patients was performed (Anonymous 2004). Six months of treatment with 400 mg of Celebrex twice a day resulted in a 28 percent reduction in polyps, compared with a reduction of 4.5 percent with a placebo (Steinbach et al. 2000). On this basis, the NDA was approved, with exclusivity for this indication expiring on December 23, 2003.

A supplement to the original NDA was filed in December 2000 to support approval for the treatment of acute pain in adults and primary dysmenorrhea, menstrual pain. This time Searle provided sufficient evidence that Celebrex relieved moderate to severe pain in patients with post–oral surgery pain, postorthopedic surgical pain, and dysmenorrhea and that a single dose of Celebrex provided pain relief within sixty minutes (CDER 2001a). The sNDA was approved in October 2001 with only three further written submissions from Searle.

However, both Searle and the FDA expended much more effort and time dealing with the issue of the effects or lack of effects of Celebrex on the gastrointestinal tract. To fully exploit the rationale behind the selectivity of COX-2 inhibitors by focusing on reduced damage to the gastrointestinal tract and associated illness and death, the approved label must support such claims. Despite the best efforts of Dr. Needleman and his colleagues, the FDA and its Arthritis Advisory Committee did not remove or dramatically modify the NSAID warning on the Celebrex label on the initial approval, though the FDA allowed the inclusion in the Special Studies Section of the results of four twelve-week endoscopic studies (CDER 1999b). Two of these studies compared the incidence of ulcers seen at 3 months in OA and RA patients taking Celebrex or naproxen. The third study provided results of monthly endoscopy in patients taking Celebrex or naproxen, and the fourth study compared monthly endoscopy results in patients taking Celebrex, diclofenac, or ibuprofen. Patients taking Celebrex were significantly less likely to have ulcers on endoscopy than those taking naproxen or ibuprofen in all comparisons. The results compared with diclofenac did not consistently favor Celebrex. The label stated the uncertainty of the correlation between the endoscopic findings and clinically significant upper GI events. In 1998, Searle/Pharmacia, in response to the recommendations of the Arthritis Advisory Committee and the FDA that clinically significant endpoints, not simply endoscopy, were required to modify the NSAID label, sponsored a trial called CLASS (Celecoxib Long-term Arthritis Safety Study). CLASS was designed to determine whether Celebrex at two and four times the maximum RA and OA doses was associated with a lower incidence of significant upper GI toxic effects compared with 800 mg of ibuprofen three times a day or 75 mg of diclofenac twice a day (Silverstein et al. 2000). As in all clinical studies, subjects would also be monitored for other adverse events. The subjects were treated and followed for six months, though the trial took nineteen months to complete, enrolling 8,059 subjects at 386 clinical centers. Published in 2000, the results were disappointing. A statistically significant difference for the primary endpoint, the incidence rates per year of upper GI ulcer complications, was not reached, though the trend favored Celebrex, 0.76 percent versus 1.45 percent. When the incidence of upper GI ulcer complications was combined with the incidence of symptomatic ulcers, the rate of 2.08 percent for Celebrex versus 3.54 percent for the NSAIDs did achieve statistical significance.

This issue of statistical significance is very important because it supports the conclusion that the differences did not occur by chance and would be seen when larger numbers of people were taking the drugs. Clinical trials, large or small, represent a small sample of the general population, and the statistical tests are used to provide confidence that the outcome did not occur by chance. CLASS subjects were allowed to take low-dose aspirin, in wide use to reduce the risk of heart attack. When the groups were divided into those also taking or not taking aspirin, better results for Celebrex were seen with both the primary and combined GI safety endpoints for only those not taking aspirin. No difference was seen for cardiovascular events (e.g., heart attack, stroke). The supplemental NDA (sNDA-09) was submitted on June 12, 2000, and then a long regulatory dance began. Meetings, teleconferences, letters, and faxes led up to an Arthritis Advisory Committee meeting on February 7, 2001.

At the same time, Merck, facing the same marketing conundrum with their COX-2 inhibitor, Vioxx, had sponsored their GI safety trial, VIGOR (Vioxx Gastrointestinal Outcomes Research). COX-2 selectivity favors Vioxx, at least theoretically, because Vioxx is four times more selective than Celebrex for COX-2 (Riendeau et al. 2001). The VIGOR trial, an international study at 301 centers, compared a high dose of 50 mg of Vioxx to 500 mg of naproxen twice a day in 8,076 patients with rheumatoid arthritis. Aspirin and other NSAIDs were not permitted. During the nine-month study, the Vioxx patients experienced ulcers and/or bleeding, obstruction, or perforation caused by an ulcer statistically less often than those receiving naproxen (Bombardier et al. 2000). The day after discussing the Celebrex submission, the Arthritis Advisory Committee discussed the Merck submissions on Vioxx labeling, including a question on the cardiovascular effects of Vioxx. Although the VIGOR study did not find a significant difference in cardiovascular events as a whole, there was a fourfold larger rate of heart attacks (0.4 percent versus 0.1 percent) with Vioxx than with naproxen (Bombardier et al. 2000).

In August 2001, a paper was published that analyzed the data from both the CLASS and VIGOR trials as well as two smaller (~1,000 patient) trials (Mukherjee et al. 2001). In addition to reporting the data on the risk of cardiovascular adverse events in the individual studies, the authors found that taking all the studies together, the yearly rate of heart attacks was higher with either Celebrex or Vioxx than the rate seen in the placebo group of 23,407 in another, unrelated study of blood pressure drugs. The difference was small—0.74 percent for Vioxx, 0.8 percent for Celebrex, and 0.52 percent in the controls—

but because the number of subjects was large, the results were statistically significant. The authors only suggest that the results raise a cautionary flag and warrant further prospective trials, but the publication of the data in the *Journal of the American Medical Association* raised some alarms, and the issue had to be addressed with warnings on the labels of the two drugs (Mukherjee et al. 2001).

The negotiations for approval of the sNDAs and modifications of the label continued. In September 2001, Searle filed a dispute resolution letter with the FDA requesting a meeting with the head of the CDER. That meeting was held in November, and the negotiations on the label continued. From the time of filing of the sNDA, there had been two acting division directors; in late 2001, a permanent director was appointed, and negotiations continued through June 2002, when the new labels for Celebrex were approved. The revised Vioxx label was approved in April 2002. Both labels include a statement on increased risk of heart attack.

The revised Celebrex labels retained the NSAID warnings and added text describing the CLASS study and the fact that no difference was seen in the rate of complicated ulcers between the Celebrex group and the combined groups taking naproxen or ibuprofen. The special studies section of the label retained the text, tables, and figures on the endoscopy studies, but both the special studies and the warnings sections of the label addressed the increased risk of complicated and symptomatic ulcers in patients taking Celebrex and aspirin. This was not the marketing tool Searle sought. One can reasonably ask why Searle chose to use such a high dose in the CLASS study. There is the concern expressed in the Arthritis Advisory Committee meetings on these drugs of the tendency for physicians to prescribe and patients to use higher than approved doses of NSAIDs, particularly if the response to treatment is unsatisfactory. This situation is referred to as dose creep. In the briefing prepared for the Arthritis Advisory Committee by the FDA gastroenterology reviewer of the CLASS study, mention is made of both the issue of dose creep and the prior accelerated approval of Celebrex at the higher dose for the FAP indication based on a risk/benefit analysis (Goldkind 2001). Therefore, in the view of the FDA, this was a safety study and the higher dose was appropriate.

Searle/Pharmacia/Pfizer has also used patents to provide protection from generic competitors. Three patents held by Searle/Pharmacia/Pfizer, covering the structure of celecoxib and related compounds and their use in the treatment of inflammation and adenomatous polyps, are listed by the FDA as submitted to cover exclusivity for NDA 020998 (celecoxib). These patents expire in 2013, 2015, and 2017. Exclusivity linked to the approval of celecoxib as a new chemical entity and the use in FAP expired at the end of 2003 and 2002, respectively, and exclusivity for the pain indications expired in October 2004 (FDA 2004a).

The Field Grows

The field of COX-2 inhibitors grew. Bextra, Searle's and thus Pfizer's second-generation COX-2 inhibitor, with 3–4 times greater COX-2 selectivity, was approved for treatment of the signs and symptoms of OA and RA and the treatment of primary dysmenorrhea on November 16, 2001 (FDA 2004a; Riendeau et al. 2001). Bextra's exclusivity as a new chemical entity expires in November 2006, and the patent covering its structure and use expires in 2015 (FDA 2004a).

Merck filed an NDA for etoricoxib (Arcoxia), their second-generation COX-2 inhibitor, with over thirteen times more COX-2 selectivity than Celebrex, in 2001 and withdrew it in March 2002 in the face of concerns about its safety (Riendeau et al. 2001). After further studies, they resubmitted the NDA in 2003 and in October 2004 received an approvable letter from the FDA, calling for further efficacy and safety data. Arcoxia has been launched in 54 countries in Europe, Latin America, and Asia.

An Explosion of Problems—A Consequence of Overreaching or Just Bad Luck?

As several other companies worked to get their newer, more selective, and ostensibly improved COX-2 inhibitors to the market, the issue of the cardiovascular safety of COX-2 inhibitors exploded. The story first hit the news in September 2004 with Merck's voluntary withdrawal of Vioxx from the market because of new information about increased rates of heart attack and stroke with the use of Vioxx in a placebo-controlled trial in the prevention of Alzheimer's disease. A firestorm in the media about the safety of Vioxx, Celebrex, Bextra, and the other COX-2 inhibitors raged, fed by the reports of elevated rates of cardiovascular problems with Bextra and Celebrex. Following the withdrawal of Vioxx, new prescriptions for Celebrex and Bextra grew, only to fall when new alerts about Bextra from Pfizer and the FDA indicated heart-related risks for patients undergoing bypass surgery (Placentra 2005). In December 2004, the NIH stopped a study testing the ability of Celebrex to reduce the return of colon polyps, and Pfizer agreed to suspend all direct-to-consumer advertising of Celebrex (Placentra 2005).

With the benefit of the acuity of hindsight, one could argue that these problems only developed as the companies did what pharmaceutical companies do, try to increase the market for their drug and possibly lengthen their exclusivity by adding indications (Drazen 2005; Solomon and Avorn 2005). But just as the pharmaceutical companies were not alone in their enthusiasm for the proposed safety benefits of the COX-2–selective drugs at the outset, academic and government scientists also were excited by the possibility that the COX-2 enzyme might play a role in conditions other than pain and arthritis. As early as 1997, reports describing the presence of COX-2 protein and messenger RNA in areas of the brain involved in Alzheimer's disease sparked the idea that the enzyme might play a role in the death of nerve cells in the disease (Jouzeau et al. 1997; Kitamura 1999). Epidemiology studies had suggested that the rate of Alzheimer's disease was lower with a history of longer duration of use of NSAIDs, though a trial of the effect of traditional NSAIDs on the deterioration of Alzheimer's patients' cognitive ability did not show any benefit (Scharf et al. 1999; Stewart et al. 1997). Also, a laboratory study with cultured cells suggested that the effects of individual NSAIDs on the reduction in production of the protein thought to be critical to the nerve destruction in Alzheimer's disease was not tied to inhibition of cyclooxygenase (Weggen et al. 2001).

The second indication that drove the development of large, long-term, placebo-controlled trials was the prevention of recurrence of polyps in the colon of individuals without the familial adenomatous polyposis. Here the target group is those who have had one or more precancerous polyps identified and removed as a consequence of screening and colonoscopy, studies recommended at regular intervals for everyone over fifty. As early as 1980, researchers were reporting that anti-inflammatory drugs, including NSAIDs, slowed or prevented the growth of animal tumors of the colon and cells from those tumors in test-tube assays (Dannenberg et al. 2005; Sinicrope and Gill 2004).

In February 2005, a three-day meeting of the Arthritis Advisory Committee and the Drug Safety and Risk Management Committee was held to address the issue of the safety and risk benefits of the COX-2 inhibitors. Also on their agenda was a discussion of the safety of the whole class of NSAIDs, not just the COX-2–selective drugs, because it was the gastrointestinal toxicity of the NSAIDs that had driven the development and approval of the COX-2–selective drugs. The hints of cardiovascular risks in the various arthritis studies of Vioxx and Celebrex were unclear because the actual number of adverse events was small and it was difficult to be sure that the results, what the Advisory Committee members referred to as "the signal," were statistically significant. But both Merck and Pfizer had embarked on a series of large, long-term, placebo-controlled randomized trials to see whether either drug would slow the development of Alzheimer's disease (Aisen et al. 2003), the recurrence of cancerous colon polyps, and in the case of Vioxx reduce the risk of prostate or colon cancer. It was findings from these trials that provided the signals that worried (Bresalier et al. 2005) and also provided media attention for an FDA safety officer, David Graham, M.D., M.P.H. Some operas have less drama than the runup to and the actual advisory committees' three-day meeting in February 2005.

The voting members assembled on February 16 numbered 32, including members of both committees supplemented with consultants both from academia and the National Institutes of Health. The meeting chair was Alastair J. J. Wood, M.D., Professor of Medicine and Pharmacology at Vanderbilt University. Over three days of meetings and more than fifteen presentations (not including forty two-minute presentations from members of the public) and tens of hours of discussion, Dr. Wood kept his calm demeanor. He adjusted the agenda to assure that the committee would learn what they needed to provide the FDA with their advice. As is true for most such meetings, the FDA structured the discussion by posing a series of questions.

For each of the approved drugs, Celebrex, Vioxx, and Bextra, the same three questions were posed:

1. Do the available data support a conclusion that X significantly increases the risk of cardiovascular events?
2. Does the overall risk versus benefit profile for X support marketing in the United States?
3. If yes, please describe the patient population(s) in which the potential benefits of X outweigh the risks and what actions you recommend that the FDA consider implementing to ensure safe use of X.

Members were also asked to comment on the role, if any, of the concomitant use of low-dose aspirin in reducing the cardiovascular risk in patients treated with the COX-2–selective NSAIDs and to propose any additional trials seen as essential for evaluating the potential cardiovascular risks and potential benefits (such as GI safety) of the three approved drugs.

The safety and labeling of the more than twenty nonselective NSAIDs approved for marketing were also addressed, given that there are no long-term controlled clinical trial data to assess the cardiovascular effects of these drugs.

The trial that triggered the withdrawal of Vioxx was the APPROVe trial, a three-year placebo-controlled trial of Vioxx (at 25 mg, the standard dose) for the prevention of recurrence of colon polyps. Although at an eighteen-month interim analysis there was a significant decrease in the number of subjects with recurrent polyps, the independent safety board for the trial found nearly double the risk in the Vioxx group of heart- and circulation-related adverse events, including edema (fluid retention), elevated blood pressure, and chronic heart failure. These were the results that led Merck to withdraw the drug from the market. It is intriguing that a small increase in the rates of edema and hypertension were noted in the safety summary of the original Vioxx NDA, but it did not appear to be of great concern to the FDA reviewers (CDER 1999c).

Five Celebrex trials received scrutiny by the advisory committee, the earliest a one-year Phase 2 trial started in 1997 to test whether Celebrex at 200 mg twice a day could slow progression of dementia in Alzheimer's disease patients. This trial did not show any efficacy. An "imbalance" in serious heart-related adverse events was seen with Celebrex, but records of the pretrial history of heart and circulation problems in the Celebrex-treated group did not allow a conclusion to be drawn about the safety of Celebrex in an elderly population (CDER 1998b). The second trial discussed was the CLASS trial, described earlier, a nine-month trial in arthritis patients that compared high doses of Celebrex with ibuprofen or diclofenac (both considered non-selective COX inhibitors, though further analyses have suggested that diclofenac is as COX-2–selective as Celebrex). As carefully phrased by the FDA medical reviewer, "the data do not support any apparent adverse cardiovascular effect of celecoxib" (CDER 2002b).

The somewhat conflicting results of two large three-year, placebo-controlled trials to test the ability of Celebrex to reduce the recurrence of polyps added to the confusion. Both trials were done in patients who had had all evident polyps removed during colonoscopy. Over 2,000 patients enrolled in the APC trial sponsored by the Division of Cancer Prevention of the National Cancer Institute (NCI) with the support of Pfizer were randomized in a 1:1:1 ratio to receive 200 mg of Celebrex twice a day, 400 mg of Celebrex twice a day, or a placebo. The PreSAP trial sponsored by Pfizer randomized 1,561 patients in a ratio of 2:3 to take either a placebo or 400 mg of Celebrex once a day. The three-year treatment periods for both studies were to be completed during 2005. Independent data safety monitoring boards (DSMBs), panels of medical and statistical experts, for each trial reviewed the data twice a year to assess whether there were safety or efficacy reasons to change the trials. At all DSMB reviews prior to December 16, 2004, the panels found no reason to change the course of the trials. Upon the withdrawal of Vioxx, the NCI requested the formation of a panel of experts on cardiovascular safety to review the cardiovascular safety data of the APC trial. Pfizer requested that this group also review the PreSAP trial data. They performed a thorough review that included revaluation of all the details of the tests done to diagnose cardiovascular problems. Using a well-established scoring system for cardiovascular adverse events, they found the risk, compared with a placebo, was more than double for both Celecoxib groups in the APC trial. But they found no difference from a placebo for the Celebrex group in the PreSAP trial. Remember that the dose and schedule were different in the two studies. The opinion of the expert committee was that continued exposure to Celebrex placed patients at risk for serious adverse events and both DSMBs recommended the trials be suspended.

Actions around another large placebo-controlled trial of Celebrex added to the drama. The ADAPT trial, sponsored by the National Institute of Aging of the NIH and run jointly by the University of Washington and The Johns Hopkins University, is a double-blind trial testing whether naproxen (220 mg twice a day) or Celebrex (200 mg twice a day) versus a placebo reduces the incidence of Alzheimer's disease (AD) in elderly subjects without dementia but at risk for AD because a first-degree relative had been diagnosed with AD. The trial began in July 2001 and by December 2004 had enrolled over 2,400 subjects, the goal being 4,500 subjects. The data safety monitoring board for the ADAPT trial met twice a year from the trial's start and at their meeting on December 10, 2004, after reviewing the data on 750 subjects who had been taking Celebrex for 1.5 years or longer, they saw no reason not to continue the study. However, when the DSMBs suspended the PreSAP and APC studies, the safety board for ADAPT suspended enrolment and treatment. Although the Pfizer briefing to the advisory committee included a summary of the ADAPT data presented at the December 10 meeting suggesting that naproxen but not Celebrex was associated with an increase in GI and cardiovascular events, Dr. Constantine Lyketsos, M.D., of Johns Hopkins University, speaking on the third day on behalf of the ADAPT steering committee, described the decision to suspend the ADAPT trial as being logistic rather than medical. He stated that the safety committee had seen at the December 10 meeting what they described as a weak signal for an increase in cardiovascular events with Celebrex, one not sufficiently worrisome or compelling to stop or change the trial. However, with the news of the suspension of the APC and PreSAP trials and the realization that one of the doses of the APC trial was identical to that used in the ADAPT trial, they realized that many current participants were likely to withdraw from the trial rather than risk taking what might be seen as a dangerous drug based on the withdrawal of Vioxx and the suspension of the ACT and PreSAP trials. Further recruitment would also be problematic. Doctor Lyketsos would not provide any data on safety or efficacy because the data were still under review (CDER 2005). One of the unintended consequences of the decision was that press reports implied that the ADAPT trial was stopped because of safety concerns about naproxen, the ingredient in the over-the-counter product Aleve.

The Bextra studies that triggered the alerts from Pfizer and label changes were trials to see whether the intravenous, water-soluble version of the active ingredient of Bextra would provide pain relief for patients who had undergone coronary bypass surgery. These studies involved a few days of intravenous dosing, when patients are unable to swallow, followed by oral Bextra. This was a placebo-controlled trial in which the patients continued to receive, as needed, the standard opioid pain medications. The problem was that the patients receiving the Pfizer drugs had an increased rate of serious cardiovascular problems (Nussmeier et al. 2005; Ott et al. 2003; Pfizer 2005).

While the presentations by company representatives and advisory consultants on drug safety and problems in drawing conclusions about the statistical significance (that is, the predictive power) of rare events were being provided, the head of Merck research, Peter S. Kim, galvanized attention, particularly of the stock analysts in the room, on the second day when he publicly commented to the meeting that if the committee concluded that the elevated risk of cardiovascular effects was a "class effect" shared by all COX-2–selective agents, Merck would consider returning Vioxx to the market (Mathews and Hensley 2005).

The discussion and voting on the questions provided a sense of the unease of the medical community about the widespread adoption of an entirely new class of drugs for non–life-threatening conditions, particularly when the duration of use will extend well beyond the duration of trials done for approval. One statistical issue brought up several times was that most of the long-term studies of these drugs were not subject to "intent to treat" analyses in which the findings with every patient who entered a trial are included, not just those who completed the trial. This can be a significant issue with drugs that have side effects that concern or bother the subjects, many of whom may drop out of the trial. Without intent-to-treat analysis, researchers lose information from those subjects, and the studies fail to provide a potential prescribing physician with what to expect if he or she prescribes a drug. Would you choose a three-year weight-loss plan by considering only those people who completed all three years? Perhaps, but you might not know that just as many people dropped out after two years because they failed to lose or gained weight. This is why intent to treat has become a standard analytic method.

Dr. Graham, of CDER's Center for Drug Safety, famous for having blown the whistle because his supervisors blocked and then permitted the publication of the epidemiologic analysis of the risk of heart attack and sudden cardiac death in patients in the Kaiser Permanente HMO system, presented his data that Vioxx but not Celebrex increased the risks and that naproxen did not reduce the risks. The comparator was previous use of an NSAID (Graham et al. 2005). Although these data added to the picture, several members of the committee stated their belief in the superior power of well-designed, randomized, and controlled prospective trials when compared with retrospective epidemiology analyses.

How did the voting go? All three approved COX-2 selective inhibitors received unanimous yes votes as significantly increasing the risk of cardiovascular events. The vote in favor of continued marketing of Celebrex was 31 to 1. For Bextra and Vioxx, the votes were positive but less decisive, 17 yes, 13 no, and 2 abstentions for Bextra, and 17 yes, 15 no for Vioxx. There was agreement that the warning of cardiovascular risk should be strengthened in language on the labels of all three drugs. Several members mentioned research that had suggested that labeling changes are not very effective in changing physician prescribing patterns and urged that the warning be a black box warning, the most compelling kind of warning on a drug's label, provided in bold font and contained in a black frame. The vote on including a warning about the lack of data on cardiovascular safety in the label for all NSAIDs was 28 to 0 (a few voting members having departed the meeting). Lengthy discussion about what kind of trial or trials are needed to clarify the issues of the relative cardiovascular safety of the NSAIDs, triggered both by the FDA's question and a suggestion by Dr. Robert Temple, Director of CDER's Office of Medical Policy, that what he called an ALLHAT trial be done to compare the cardiovascular effects of NSAIDs using naproxen and diclofenac as controls. Whether such a megatrial could be done and who would fund it remained unclear, though the enthusiasm among the members of the committee was high.

On April 7, 2005, the FDA and Pfizer both announced that the FDA had requested that Bextra be removed from the market and that Pfizer had complied. The FDA's decision on Bextra was driven not only by the lack of data on cardiovascular safety but also the risk of serious, life-threatening skin reactions (FDA 2005). The FDA also asked Pfizer and all of the manufacturers of all NSAIDs, COX-2–selective or not, to include on their label a boxed warning "highlighting the potential for increased risk of cardiovascular (CV) events and the well described, serious, potential life-threatening gastrointestinal (GI) bleeding associated with their use" (FDA 2005). The company stated in their press release that they respectfully disagreed with the FDA's decision on Bextra and would be working with the FDA on the language of the Celebrex label (Berenson 2005). The financial consequences to Pfizer are unclear. They now have with Celebrex, black box warning notwithstanding, the only COX-2–selective agent currently on the U.S. market, providing $3.3 billion in sales in 2004, compared with $1.29 billion in sales of Bextra (Mathews and Hensley 2005a, b). What will happen to Merck's Vioxx, or their Arcoxia, now sold in 54 countries, and Novartis' COX-2–selective agent, Prexige, is unclear (Mathews and Hensley 2005a, b).

PATENTS: ALWAYS AN ISSUE

Searle was involved in two patent disputes that provide insight into some of the legal struggles engaged in by companies over blockbuster drugs. In 1997, in Europe Pharmacia filed suit against Merck, alleging that Vioxx infringed on the European Celebrex patent. The suit was eventually dismissed (Pharma Marketletter 2002). Another patent battle was launched by the University of Rochester, which held the rights to a patent # 6,048,850, filed in 1995 and issued in April

2000, claiming the gene for the enzyme they called prostaglandin H synthase-2 (i.e., COX-2) for the discovery of specific inhibitors and the use of such inhibitors to treat inflammation. The University was claiming that Pharmacia had infringed on its patent in the discovery of its COX-2 inhibitors and sought payment equal to 10 percent of the sales of Celebrex (Ebert 2000). The legal issues are complex and hinge on the precise use of words in the university's patent, but in the end the decisions of the trial court and the appeals courts in dismissing the case were based on the evidence that the patent included no claim for a compound discovered by the method (*University of Rochester v. G.D. Searle & Co., Inc.* 2003).

The FDA often obtains commitments from sponsors to complete postmarketing studies to answer questions regarding safety, efficacy, and how best to use the drug but do not, in the opinion of the FDA and its advisors, prevent the approval of the drug (CDER 2004a). These Phase 4 study commitments are published on CDER's Web site, and sponsors are required to report each year on their progress in completing the commitments (CDER 2004a). Teva was required to complete two Phase 4 studies: to determine whether patients taking Copaxone developed complexes of antibody and Copaxone that adversely affect immune function; and to evaluate whether Copaxone caused modification of immune responses to antigens normally eliciting a cellular immune response. These studies were completed and submitted to the FDA.

When Searle received accelerated approval for Celebrex to treat FAP, the FDA required a commitment to complete a randomized controlled trial on familial adenomatous polyposis (FAP) to verify and describe the clinical benefit of Celebrex in this population and also to establish a long-term registry of clinical outcomes in FAP patients.

These two drugs provide a snapshot into two ways an innovative drug is discovered and moves into the armamentarium of physicians treating serious diseases. One, Copaxone, began life relatively modestly as a way to test new ideas of how novel molecules could be used to probe the immune system. First tested in humans in the 1970s, it took decades of persistence on the part of its inventors and a pioneering clinical investigator who did not live to see its approval. Copaxone's market share in June 2003 was 26–27 percent in the United

States and about 12 percent in Europe, numbers seen as less suscepti-ble than Avonex's to patients' anticipated switch to Rebif (Mehrotra et al. 2003). Worldwide sales of Copaxone for 2006 are projected as $865 million (Mehrotra et al. 2003). Research on the potential utility of Copaxone in animal models of neurodegenerative disorders such as Amyotrophic Lateral Sclerosis (ALS) by Professor Sela and colleagues is promising, so there may be other small, niche markets for the drug (Angelov et al. 2003).

Celebrex is another story. Its discovery was purposeful, its develop-ment intense, and its marketing masterful. It achieved a billion dollars in sales in its first years, in a mature market with many older compet-ing drugs with which physicians were comfortable and with other similarly targeted competitive drugs not far behind in the pipeline. Being first to market had its charm and, even in 2003, Celebrex showed only slowly declining sales. According to a CBS report based on data from Verispan, a company that tracks prescriptions, Celebrex sales grew 75 percent in 2004 to $3.3 billion. During the week of De-cember 17, 2004, Celebrex had a 44 percent share of the prescrip-tion pain relief market, earning $44 million, but by February 11, 2005, the share had fallen to 23 percent, with sales of $24 million (CBSnews.com 2005). Just a reminder that human biology remains a daunting master, and no one could have predicted what the field of COX-2 inhibitors would become and what challenges would arise.

The companies developing the two drugs also differ. Teva was a generic powerhouse and remains so, with an interest in getting into innovative drugs and a compound with promising clinical data at a nearby research powerhouse. Philip Needleman joined Searle/Mon-santo with the discovery of a COX-2 inhibitor as a major goal and the people and resources to make it happen. The timing was right. The science was there. And many large companies had their sights set on the profitable inflammation market, filled with drugs with expir-ing patents and side effects thought by the companies as more and more troubling to physicians and regulators. With the merger and ac-quisition fever in the pharmaceutical industry, Searle's drugs ended up in the Pfizer machine, now the world's largest pharmaceutical company, with the benchmark sales and marketing force. So, until the fall 2004, we had the appealing TV ads with lithe seniors running on the beach, doing Tai Chi, or just walking the dog, all (including the dog, who also can get an NSAID for his painful joints) smiling without pain. Those ads may never return, but others have taken their place and, as we speak, may be poised to put their champions on the medicine rack.

18

Are We (Well) Served? Do We Have the System, Industry, and Regulations We Need, Want, and Deserve?

THE INDUSTRY

Do we have the system we need for the discovery, development, review, and provision of new medicines? The discovery, development, and provision aspects are, perhaps not unreasonably for a capitalist society, big business, employing over 400,000 people, expending $33.2 billion on R&D, and generating $230 billion in sales in the United States and Canada, $154 billion in the United States (Gale 2005; PhRMA 2004). A 2004 report from the Milken Institute states that the biopharmaceuticals industry was responsible for $63.9 billion in real output in 2003 without factoring in the ripple effects on other sectors of the economy, an exercise that generated a total of $172.7 billion in output (DeVol 2004). So the industry is good for the economy. That is not the whole story. Despite Charles Wilson's 1951 comment, "What's good for the country is good for General Motors, and vice versa," economic success may not be the sole or dominant criterion on which to judge a company or an industry. We are ambivalent about the health care industry. Americans may embrace a utilitarian view for most commercial activities, but health care is often viewed as a communitarian right. We want the substantial return to our 401Ks of our investment in drug company stocks. We also brook at the withholding of medical care based on the ability or inability to pay, yet many resist the implicit transfer of resources by a proposed increase in the FICA medical deductions from our paychecks. Just as the ingestion of a powerful drug comes at the risk of side effects that may counterbalance the benefits, there may be no economic free lunch in the area of health care in general and prescription drugs specifically. That is the argument of the pharmaceutical companies and their lobbying organization, PhRMA: the prices and profits are what fund the R&D for the next drugs (PhRMA 2004).

Since the 1980s, the pharmaceutical industry has gone through dramatic changes as the drive for greater profits and market share in the face of expiring patents has sparked many megamergers, creating international companies with sales in the tens of billions of dollars and market valuations in the hundreds of billions. The legalization of direct-to-consumer marketing and the increasing investment in research as high-technology discoveries and tools seemed to demand retooling of the discovery enterprise had led to a focus on blockbusters. The out-of-the-box success for the launch of Celebrex (to say nothing of Viagra et al.) has reinforced the idea that only a blockbuster, a drug bringing in $300 million in sales by the third year of marketing, is worth the investment. This means that the focus within many of the large and ever-growing biopharmaceutical companies has shifted in a subtle way from medical needs to markets. I guess we are lucky in the developed world that improved public health means that we get sick and die from a shrinking number of serious diseases (heart disease and cancer) that allow for blockbusters, and many of us have the cash or the insurance to gain treatments for the less deadly and even just the annoying medical conditions.

Are we well-served? The question begs for numbers. Life expectancy in the United States lengthened during the twentieth century from 47 to 77 years (CDC 2004). A simple comparison of the leading causes of death in 1900 and 1998, tempered by the caveat that the classifications have changed with the times, tells a story (CDC 2000). In 1900, the top three causes of death were (1) pneumonia and influenza, (2) tuberculosis, and (3) diarrhea, enteritis, and ulceration of the intestines. Together, these three causes accounted for nearly one-third of recorded deaths. In 1998, pneumonia and influenza, the only infections in the top ten, ranked sixth, accounting for fewer than

4 percent of deaths. Diseases of the heart, cancers, and strokes were the top three, accounting for over 60 percent of recorded deaths. The widespread availability of vaccines and antibiotics may account for some of these changes, but public health measures, including access to clean water and improved sanitation (indoor plumbing and attention to the hazards of exposure to sewage), must be given credit for the 90 percent drop in infant mortality and 99 percent drop in maternal mortality that account for a great deal of the increase in life expectancy (MMWR 1999).

Another story that deserves telling is the impact of drug therapy on those who become infected with the human immunodeficiency virus-1 (HIV-1). The alarm was first raised in May 1981 by the U.S. Centers for Disease Control (CDC) that men were becoming ill and dying from a rare form of pneumonia. The causative agent for the destruction of the immune system that led to those infections and deaths, HIV-1, was isolated in 1984 by French and American researchers, a situation that led to an unseemly battle for credit and money because the rights to a blood test were at stake. Zidovudine (AZT), the first drug to treat the virus, was approved in March 1987, but single-drug therapy had limited benefit, and other drugs with a similar mechanism of action added little. In 1997, the first of an entirely new class of drugs, the protease inhibitors, was approved, ushering in the use of a cocktail of three or more drugs, collectively known as HAART (highly effective antiretroviral therapy) (Hoffmann and Kamps 2003). The truth of the HAART label can be seen by a glance at a few numbers describing the proportion of individuals in developed countries that are alive ten years after becoming infected with HIV, as detected by a positive test for antibody to the virus (CASCADE Collaboration 2003). An analysis of patients from Europe, Australia, and Canada uncovered that 64 percent of 16–24-year olds and 28 percent of those over 45 who became infected with HIV-1 before 1997 were alive at ten years. For the period between 1999 and 2001, the projected numbers were 94 percent and 87 percent for the two age groups. The FDA as of October 2004 had approved twenty-six drugs or drug combinations to control the reproduction of the virus and control the progression of the disease. These drugs are not cures, but they have converted a relentless and speedy killer into something approaching a chronic disease. The use of one of the antiviral drugs during the second and third trimester and at delivery has dramatically dropped the rate of mother-to-child HIV transmission in the United States (Santora 2005). The treatments are not without side effects, and the increased rate of resistance to the drugs has slowed the rate of improvement in survival (CASCADE Collaboration 2003). There are also the issues of cost and access. The literature on the issue of inequality of access to HIV drugs is large and the debates rage, but equitable or not, the prospects for an individual infected with HIV in developed countries have changed dramatically, in large part because of the R&D efforts of large pharmaceutical companies. Yes, some of these drugs were first discovered at universities and at the NIH, but without the organized effort of the pharmaceutical companies, the drugs would not have been manufactured, tested, and approved for sale.

Can a similar story be told for cancer? Drs. Paul Calabresi and Bruce A. Chabner, pioneers in the field of medical (as distinct from surgical) oncology, have written that "curative treatments have been identified for a number of previously fatal malignancies such as testicular cancer, lymphomas, and leukemia" (Calabresi and Chabner 1996). This is progress, even if the cancers listed are relatively rare. The situation for more common cancers is less dramatic, but five-year survival rates for the more common cancers have risen (ACS 2004). Five-year survival for breast cancer was 50 percent in 1974–1976 and 87 percent in the period 1992 to 1999, in large part because of the development of paclitaxel (Taxol) and related drugs. The survival rate for prostate cancer rose from 67 percent to 98 percent during the same periods.

Between 1950 and 2003, the FDA approved sixty new cancer drugs (CDER 2004b). Two successes cited by Calabresi and Chabner, leukemia and testicular cancer, deserve some exploration. The results for childhood leukemia, specifically the type called acute lymphoblastic leukemia (ALL), are indeed a triumph. The improved five-year survival of children with acute lymphoblastic leukemia, from 21 percent in 1961–1966 to 86 percent in 1997, has been attributed to the use of combinations of drugs, the earlier use of chemotherapy introduced directly into the central nervous system (the brain and spinal cord) in recognition of the failure of drugs to pass from the blood into the central nervous system, and the use of higher doses of drugs for longer periods of time. The drugs that form the basis for most regimens in childhood ALL are methotrexate, dexamethasone, and vincristine. Lederle Labs gained approval for methotrexate in 1953, Merck gained approval for dexamethasone in 1958, and Lilly began marketing vincristine in 1963. The improvement in survival of children diagnosed with ALL was in large part the result of clinical trials organized and run by cooperative groups of physicians and research institutes and funded by the National Cancer Institute of the NIH. But the cooperative groups and hospitals had to have the drugs to test.

Ninety percent of men with certain forms of testicular cancer are cured with a regimen that includes three drugs by Bristol-Myers Squibb: cisplatin, approved in 1978; bleomycin, approved in 1973; and etoposide, approved in the United States in 1983 (Bristol-Myers Squibb). This represents a dramatic improvement for this rare but previously deadly disease (Loehrer and Einhorn 1984). Cisplatin was the result of the work of an academic, Barnett Rosenberg, who while studying the effects of electric currents on the ability of bacteria and other cells to divide discovered that the interference was caused by leaching of a form of platinum from the platinum electrodes. After animal tests at the University of Michigan and confirmation at the NCI, cisplatin entered human trials at the NCI and at cancer centers around the United States and Europe, supported in large part by the NCI. Changing the way the cisplatin was given solved the problem of severe kidney toxicity. Less than ten years elapsed between the discovery of the effects of cisplatin and its approval, and although taxpayer funds contributed to the studies, licensing of the compound to a pharmaceutical company, Bristol-Myers, was critical for its approval and availability (Nader and Love 1993). Bleomycin, an antibiotic anticancer drug discovered by a Dr. Umezawa of the Tokyo Institute for Microbiological Chemistry, was also licensed by Bristol, as was etoposide, a derivative of a mayapple toxin first synthesized by Sandoz Laboratories, now part of Novartis (Landau et al. 1999). Bristol Labs, now Bristol-Myers Squibb, was in the 1970s to 1990s focused on becoming the leader in cancer treatments, and their subsequent licensing of paclitaxel (Taxol) from the NCI cemented this position, which began to wane as other large pharmaceutical companies entered the field. It could be argued that the clumsy deal with ImClone for Erbitux was part of their struggle to regain this position. But that is another issue for another venue.

The results for lymphoma, particularly non-Hodgkin's lymphoma, also are based on combinations of radiation and chemotherapeutic drugs discovered and approved in the 1970s to late 1980s.

Clearly, these life-saving drugs were only going to reach patients not enrolled in clinical trials if an NDA were submitted to the FDA and approved, with all the expense that entails, something universities and government agencies are not set up to do. But this argument begs the question. Private industry delivers the life-saving drugs because the system for the provision of drugs that we have established places that role in the private sector. Could we do better?

No developed country relies on the government or public sector to discover, develop, and test drugs. Let us consider Cuba, a country that meets its citizens' health care needs, including medicines, ostensibly without private industry. Cuba does indeed provide the basic drugs through government-run labs and research institutes, but outside of the area of certain biotechnology products where Cuban researchers have discovered innovative drugs that are exported to bring in hard currency through business collaborations, the Cuban system, hit with the collapse of the Soviet Union and the U.S. economic embargo, has focused on providing the essential medicines, manufactured internally from imported raw materials (La Porte 1997; SPP 2004). With a large number of well-trained physicians and an emphasis on preventive care, maternal and child health, and communicable diseases, Cuba boasts infant mortality and life expectancy numbers comparable to developed countries and much better than the rest of Latin America (SPP 2004). This success may challenge the idea that a private pharmaceutical industry is required to provide health care, but the drugs that the Cuban system manufactures for internal use and for export are, with the exception of the biotechnology drugs and vaccines discovered and developed by its research institutes with substantial Soviet subsidies, copies of innovator drugs from the developed world (Tancer 1995). Cuba is thus a straw man for an argument against the U.S. prescription drug industry. The problem is that there are no relevant data to say it could be better if organized differently. But we do pretty well, if you are part of the "we" with health insurance that includes prescription drug coverage.

REIMPORTATION

Over the last decade the media has focused a great deal of attention, with a reciprocating regulatory clucking by the FDA, on the growing practice of reimportation of approved pharmaceuticals from Canada that were manufactured in a registered facility (Choudhry and Detsky 2005). Often these drugs were manufactured in the United States but cost less in Canada, so individuals and groups of patients, often senior citizens on fixed incomes and with little or no prescription drug health insurance coverage, travel to Canada or use the services of Canadian online pharmacies to purchase their prescribed drugs. The news stories often point to savings achieved by these individuals as evidence of our government's failure to control, or even its complicity with, rapacious drug companies. The political use of this issue has been healthy and steady on all sides and was intensified with the passage of the Medicare Prescription Drug, Improvement, and Modernization Act of 2003, which contained language that precluded Medicare from using its buying power to negotiate with manufacturers for lower prices. The crafted language is precisely that, "In order to promote competition under this part and in carrying out this part, the Secretary [of HHS] may not interfere with the negotiations between drug manufacturers and pharmacies and Part D sponsors" (insurance and managed care organizations) (Medicare Prescription Drug, Improvement, and Modernization Act of 2003). Such negotiations and bargaining by federal agencies is not illegal; the Veterans Administration uses its buying power to negotiate discounts for drugs (Katz and Deshpande 2005).

Reimportation of drugs is possible within the law, but reimporters must meet high standards for accountability. The U.S. Code requires that there be a statement provided for such reimported drugs that "identifies the manufacturer of such article and each processor, packer, distributor, or other entity that had possession of the article in the chain of possession of the article from the manufacturer to such importer of the article" (USC). The FDA's position is that without documentation of the chain of custody and evidence that at each stage of its travels the required testing was performed to assure that what is stated on the label is what is in the bottle, American patients are at risk. Few commentators question the professionalism and quality of Canadian pharmacies, distributors, and wholesalers, but as any regulatory entity will claim, the FDA's mantra is that if it is not documented, it was not done. Guilty until proven innocent. This may seem harsh, but it is the FDA's job to protect us from adulterated drugs, which history has shown can be dangerous and will proliferate unless controlled.

What often is not explored too deeply is why some, but not all, drugs are less expensive in Canada. Canada has a universal health care system, partially funded by the federal government, with ten different provincial systems. Drugs are not covered by the federal system, but provincial governments may provide reimbursements for medicines for senior citizens and those receiving social assistance. Many Canadians also carry private health insurance, and many of these companies have developed cost-control schemes similar to those in the United States, providing for lower out-of-pocket expenses for generic drugs or those included in a formulary (Elgie 1996). The Canadian government has both a limited system of compulsory licensing of patent-protected drugs to generic manufacturers and a quasi-judicial agency, the Patent Medicine Prices Review Board, that controls excessive prices. They do this by reviewing information provided to them by the manufacturer, starting when a drug is first approved for sale by Health Canada (Canada's counterpart to the FDA) and at six-month intervals as long as the patent is in force. The board works on three principles:

- The prices of most new drugs should not exceed the prices of drugs currently on the market that treat the same disease.
- On average, prices in Canada should not exceed the median prices in other industrialized countries.
- Prices should not go up faster than the Consumer Price Index.

The board has the authority, if they find prices exceed the guidelines, to investigate and require not only conforming price reductions but also punitive reductions and fines to offset the excessive revenues. With these rules in place, the board assures that Canadians do not pay more than the median of foreign prices for the top-selling patented drugs.

In addition to the news reports, several analyses have supported the claims that prescription drug prices are higher in the United States than in Canada or the United Kingdom (GAO 1992, 1994). It is important when faced with such comparisons that we consider not just the prices for a small number of widely known and advertised blockbuster drugs but also the range of available drugs. A 2003 analysis of prices and availability of pharmaceuticals in nine industrial countries found that, as expected, prices in the United States for on-patent, brand-name drugs were higher in the United States than in Canada, Chile, France, Germany, Italy, Mexico, and the United Kingdom, though 40 percent lower than in Japan. However, the prices for generic drugs were lower in the United States than in Germany, Italy, Japan, Mexico, and the United Kingdom and comparable to the prices for generics in Canada and Chile (Danzon and Furukawa 2003). In the United States, generics account for a larger percentage of the market for prescription drugs sold, either by volume or sales. These price comparisons matched the drugs in active ingredient and dose. Over-the-counter drugs were more expensive in every country studied than they were in the United States. The authors argue that exchange rates and price-control policies contribute to these differences in price and that fewer new, innovator drugs are available and in use outside of the United States (Danzon and Furukawa 2003). The first point may shift with changing exchange rates for the dollar, and the second point begs the question as to whether we are better served by the newer drugs. The stories of the drugs we are tracking suggest a mixed answer. Patients with MS have benefited from an alternative to the interferons to slow the ravages of their disease. In large numbers, people have benefited from improved relief from arthritis pain with Celebrex and arguably the other COX-2–selective drugs, even in the face of a statistically significant though small increase in the risk of cardiovascular problems. Many achieve pain relief without the larger risk of serious GI problems. Only as the FDA moves to strengthen labeling will we be able to see whether patients and their physicians judge that the risk-to-benefit ratio favors the use of the COX-2–selective drugs.

THE REGULATORS

The other element in the system for providing our medicines is the FDA, an agency whose rules and regulations have evolved in a punctuated fashion driven by crises and scandals, from the ethylene glycol poisonings through the thalidomide devastation to the flap over the safety of COX-2 inhibitors, a process influenced by counterbalancing exhortations to get drugs approved faster but not compromise safety. These pendulum swings in pressures may typify the regulation of any free-enterprise undertaking, but the pitch of the demands for safer and faster seems heightened by the fact that we are talking about our health, about life and death. Despite the alarms of the nearly seasonal books asserting or just hinting that the FDA and the drug industry are too cozy, so that the companies are free to deceive and harm us in their quest for profits (Angell 2004; Avorn 2004), we and the industry seem to survive and muddle through. This season's theater about the overprescribing, overmarketing, and failure to respond adequately to safety warnings of the selective COX-2 inhibitors, complete with villains (Merck management comes to mind) and heroes (think David Graham of the FDA's Office of Drug Safety), led to another round of congressional hearings and tinkering, albeit possibly useful tinkering, with the FDA regulations (Graham et al. 2005; Solomon et al. 2005). Herper, in January 2005, suggested five ways to fix the FDA (Herper 2005):

- increase the agency's funding so that it has the staff and resources to track safety after approval;
- provide the agency with the incentives and penalties to assure safety studies are done, including deferring direct-to-consumer advertising until more safety data are available and blocking approval of a generic competitor in exchange for the performance of safety studies;
- establish mandatory registries for tracking side effects;

- provide government money, perhaps through the NIH, for postmarketing safety studies; and
- develop more effective methods than label changes to gain the attention of physicians and patients when safety issues arise, perhaps with a dedicated advertising campaign.

On February 15, 2005, prior to the COX-2 Advisory Committee meeting, the secretary of HHS and then acting FDA commissioner, Lester Crawford, announced a "new, emboldened vision for FDA" that included the formation of an independent Drug Safety Oversight Board (DSB) to oversee the management of drug safety within CDER. The members of the DSB will include FDA staff and medical experts from other government agencies and departments. Transparency of the FDA's decision making to the public will also be improved (FDA 2005b).

The FDA, as with many regulatory agencies, responds to crises and crises of confidence with adjustments and tweaking of their rules, within their authority to do so, but sometimes laws need to be changed. The issue of the slow pace of labeling changes, brought about by often-protracted negotiations between the FDA and the drug manufacturer, was addressed on March 1, 2005, when Sandra L. Kweder, the FDA's deputy director for new drugs, asked a Senate committee for the authority to dictate label changes (AP 2005; U.S. Senate 2005).

A second issue highlighted by the discussions and votes of the COX-2 Advisory Committee meeting was the regulations controlling direct-to-consumer advertising. The committee members voted unanimously to prohibit DTC advertising for the COX-2 inhibitors and prescription NSAIDs in general. Yet this is not entirely under the control of the FDA. What may be said and what must be said in such ads are within the FDA's purview and must be based on the label with its warnings, but the freedom to publish and broadcast such ads is in a sense a free-speech issue subject to evolving case law and jurisdictional deliberation between the Federal Trade Commission and the FDA. Whether Congress and the president seek to change the rules surrounding DTC advertising is not known at present. PhRMA is developing voluntary guidelines and at least one company, Bristol-Myers Squibb, has pledged to forego DTC for one year after approval of a new drug.

Another recurring issue surfaced a few days after the Joint Advisory Committee meeting on the COX-2 inhibitors, when the *New York Times* reported the results of analysis done at the paper's request by the advocacy group the Center for Science in the Public Interest. The Center tabulated the ties of the voting members with the companies whose drugs were under review and the impact of those with such ties on the votes. According to this analysis, and CSPI provided names and details, ten voting members were consultants and/or speakers for one or more of the companies marketing or developing COX-2 inhibitors (CSPI 2005). The *Times* reported that if these ten had not voted, the committee would have voted 12 to 8 to withdraw Bextra and 14 to 8 to withdraw Vioxx. The *Times* interviewed eight of the ten, and all indicated that their past relationships did not influence their votes. The problem is that, given the increasing reliance of university biomedical researchers on pharmaceutical funding for clinical research, exclusion of every individual with ties to a company selling or developing an NSAID would have precluded formation of a knowledgeable panel. The companies go to these folks because they are the experts. So the FDA employs a system to review the personal and professional financial ties of each voting member for each meeting and excludes participation or voting by those who would benefit from the outcome of a vote on a particular product. Because the joint committee was considering many drugs from many different companies, the FDA's judgment was that the discussion involved "issues of broad applicability and there are no products being approved." That is not quite true because the committee did vote in response to the FDA's questions on whether they thought the three approved drugs should be withdrawn. So in such a controversial situation, the agency might have been wiser to acknowledge and disclose anything that might be perceived as a conflict of interest and have the appropriate members recuse themselves. Perhaps next time.

The April 2005 decision by the FDA regarding the availability of Bextra and the labeling of the COX-2 selective and nonselective NSAIDs clearly signaled, despite the disclaimers by the agency spokespersons of case-by-case, balanced reviews of benefits and risks of all applications, a move toward a more cautious stance (Mathews 2005). But that may last only until the next medical headline or the next news conference demanding to know why a life-saving drug available in Europe or elsewhere is not available to U.S. patients.

Beyond these adjustments and rule changes, if we ask whether we have the regulatory control for prescription drugs that we need and want, the responses from commentators will differ. Some, such as Avorn and Angell, see an unhealthy and dangerous relationship between the companies and the FDA, but others, such as Hilts, see an agency that is "not brilliant in all respects" but whose employees go to work with a sense of purpose and more often than not succeed (Angell 2004; Avorn 2004; Hilts 2003). Faint praise, but for a challenging arena such as regulating the instillation of potent stuffs into millions of bodies each and every day, the successes outnumber the lapses. Perhaps that is as good as it can be.

DO WE KNOW WHAT WE NEED TO KNOW?

One result of the growing discomfort about worrisome surprises about Vioxx, Celebrex, and other approved drugs was the public hand-wringing by journal editors such as Marcia Angell, former editor of the *New England Journal of Medicine*, and Catherine DeAngelis of the *Journal of the American Medical Association* when they began to understand that the clinical results they were publishing in their prestigious, peer-reviewed journals did not contain all the available information about safety and efficacy or fully disclose changes from the original design of the trials between initiation and completion and analysis (Angell 2004; Chan et al. 2004; Mathews 2005). The calls went out for the development of databases of clinical trials registered as they began, as well as posting all of the details of efficacy and safety results and trial design changes (Mathews 2005; Chan et al. 2004). The malefactors were not just the drug companies but the academic research physicians whose names graced the papers because they were investigators on trials that may or may not have been drug-company sponsored. The term transparency was used because analyses indicated that the physician or patient readers were not being told everything. Some of the criticisms focused on the fact that statistically significant results were more likely to be reported than statistically insignificant observations (Chan et al. 2004). While the industry and industry groups such as PhRMA and journal editors work to come up with trial registries, protocols, and complete results, the effort seems a bit quixotic. Who is going to read and judge all those files? Busy physicians who demonstrably do not always read package inserts, rely on detail persons for their information, and accede to patient requests for the newest drug featured in a TV ad? Patients, few of whom would understand the arcane language they are reading and know how to make a judgment about the lines and lines of data? There is a reason most new and powerful medicines are only available by prescription. Legally, the physician, the learned intermediary, is often and rightfully held responsible for a treatment decision. One group that might be hurt by the trial databases and disclosure would be lawyers who specialize in class action lawsuits at the first suggestion that a drug is not as safe or effective as we first believed. We, the patients and physicians, could not say we did not know. It was right there in the published protocol and results. In this writer's opinion, responsibility must be placed at the door of the prescribing learned intermediary to make sure he or she knows what is to be known before choosing to prescribe a drug, patient entreaty or not. Saying no is still possible. We as patients (an interesting term) must work to be sure that our health care system provides the physician with the tools and time to keep up to date and then empowers the physician and holds him or her responsible for keeping current. The system we have to provide and evaluate drugs, as imperfect as it may be, is what our free enterprise system provides, and although the system may need tweaking of the sort that will happen as a result of the alarms about the COX-2 inhibitors, we are doing pretty well.

Bibliography

Abraham J, Reed T. Reshaping the carcinogenic risk assessment of medicines: international harmonisation for drug safety, industry/regulator efficiency or both? *Soc Sci Med*. 2003; 57(2):195–204.

Abramsky O, Teitelbaum D, Arnon R. Effect of a synthetic polypeptide (Cop 1) on patients with multiple sclerosis and with acute disseminated encephalomyelitis. *J Neurol Sci*. 1977; 31:433–438.

Advisory Committee on Human Radiation Experiments (ACHRE). ACHRE report: the development of human subject research policy at DHEW, Part 1, Chapter 3. U.S. Government Human Radiation Interagency Working Group (IAWG). 1997 [accessed July 11, 2002]. Available from http://tis.eh.doe.gov/ohre/roadmap/achre/chap3_2.html.

Aisen PS, Schafer KA, Grundman M, Pfeiffer E, Sano M, Davis KL, Farlow MR, Jin S, Thomas RG, Thal LJ. Effects of rofecoxib or naproxen vs placebo on Alzheimer disease progression: a randomized controlled trial. *JAMA*. 2003; 289(21):2819–2826.

American Cancer Society (ACS). Cancer statistics presentation 2004. American Cancer Society. 2004 [accessed August 1, 2005]. Available from http://www.cancer.org/docroot/pro/content/pro_1_1_Cancer_Statistics_2004_presentation.asp.

American College of Physicians (ACP). Physicians and the pharmaceutical industry. *Ann Intern Med*. 1990; 112:624–626.

American Journal of Medicine. Nonsteroidal anti-flammatory drug-induced gastrointestinal damage: Current insights into patient management. Washington, DC: Symposium proceedings; June 13, 1987; 84(2A):1–52.

American Medical Association (AMA). Council on Medical Affairs: gifts to physicians from industry. *JAMA*. 1991; 265:501.

American Society of Hospital Pharmacists (ASHP). Statement of formula system. Am J Hosp Pharm. 1983; 40:1384–1385.

Anderson S. E-mail query, September 29, 2003.

Angell M. *The Truth About Drug Companies: How They Deceive Us and What to Do About It*. New York: Random House; 2004.

Angelov DN, Waibel S, Guntinas-Lichius O, Lenzen M, Neiss WF, Tomov TL, Yoles E, Kipnis J, Schori H, Reuter A, Ludolph A, Schwartz M. Therapeutic vaccine for acute and chronic motor neuron diseases: implications for amyotrophic lateral sclerosis. *Proc Natl Acad Sci USA*. 2003; 100(8):4790–4795.

Animal Welfare Act as amended. 2002; (7USC) 2131–2156.

Anonymous. Business briefs. *The Jerusalem Post*. March 9, 1995; 8.

———. Carey J. McClellen's friendlier, speedier FDA. *Business Week*. June 16, 2003:33.

———. ImClone application "scientifically incomplete" FDA states in refusal to file letter on C225. *Cancer Lett Interactive*. Special Edition, March 8, 2002; 1–8.

———. The quality function: a pharmaceutical manufacturing white paper. Best Practices, LLC. 2004.

———. 10 direct-to-consumer drugs by sales. *Advertsing Age*. December 22, 2003.

AP. FDA asks Congress for authority to dictate drug-warning labels. *Wall Street Journal*. March 1, 2005.

APHIS. Recent News of the Birds, Rats and Mice Issue. USDA, November 24, 2002 [accessed March 8, 2003]. Available from http://www.aphis.usda.gov/acrmb update.html.

Arnold, M. Direct delivers: 5 factors driving direct marketing. Medical Marketing and Media. June 1, 2003. Available from http://www.findarticles.com/p/articles/mi_hb3272/is_200306/ai_n7972227.

Arthritis Foundation. Disease Center: Osteoarthritis 2004 [accessed July 31, 2005]. Available from http://www.arthritis.org/conditions/DiseaseCenter/oa.asp.

———. Disease Center: Rheumatoid Arthritis [accessed July 31, 2005]. Available from http://www.arthritis.org/conditions/DiseaseCenter/ra.asp.

Aspirin Foundation. Who we are. [Web page] 2005 [accessed July]. Available from http://www.aspirin.org/who.htm.

Avorn J. *Powerful Medicines: The Benefits, Risks, and Costs of Prescription Drugs*. New York: Alfred A. Knopf; 2004.

Bailar JC III. The powerful placebo and the Wizard of Oz. *N Engl J Med*. 2001; 344(21):1630–1632.

Baselga J. Why the epidermal growth factor receptor? the rationale for cancer therapy. *Oncologist*. 2002; 7(90004):2–8.

Beauchamp TL, Walters L, eds. *Contemporary Issues in Bioethics*, Fourth ed. Belmont, CA: Wadsworth Publishing Company; 1994.

Beecher HK. Ethics and clinical research. *N Engl J Med*. 1966; 274(24):1354–1360.

Begley R. FDA user fees proposal is finding wide support. *Chemical Week*. August 19, 1992: 12.

Berenson, Alex. Pfizer loses one remedy for its slump. *New York Times*. April 8, 2005:Section C.

Bianchi M. Are all NSAIDs other than 'coxibs' really equal? *Trends Pharmacol Sci*. 2004; 25(1):6–7.

Biogen-IDEC/Elan. Tysabri product information. 2004 [accessed December 10, 2004]. Available from http://www.tysabri.com/downloads/product_information.pdf.

Bishop LJ, Nolen AL. Animals in research and education: ethical issues. *Kennedy Inst Ethics J*. 2001; 11(1):91–112.

Bjorkman D. Nonsteroidal anti-inflammatory drug-associated toxicity of the liver, lower gastrointestinal tract, and esophagus. *Am J Med*. 1998; 105(5, suppl 1): 17S–21S.

Bloomberg. F.D.A. to review ImClone cancer drug. *New York Times*. October 11, 2003: Page 3; Column 1.

Bombardier C, Laine L, Reicin A, Shapiro D, Burgos-Vargas R, Davis B, Day R, Bosi Ferraz M, Hawkey CJ, Hochberg MC, Kvien TK, Schnitzer TJ, The VIGOR Study Group. Comparison of upper gastrointestinal toxicity of rofecoxib and naproxen in patients with rheumatoid arthritis. *N Engl J Med*. 2000; 343(21): 1520–1528.

Bornstein MB, Miller AE, Teitelbaum D, Arnon R, Sela M. Treatment of multiple sclerosis with a synthetic polypeptide: preliminary results. *Trans Am Neurol Assoc*. 1980; 105:348–350.

Bornstein MB, Miller A, Slagle S, Weitzman M, Crystal H, Drexler E, Keilson M, Merriam M, Wassetheil-Smoller S, Spada V, et al. A pilot trial of COP 1 in exacerbating-remitting multiple sclerosis. *N Engl J Med*. 1987; 317:408–414.

Bosl GJ, Motzer RJ. Testicular germ-cell cancer. *N Engl J Med*. 1997; 337(4):242–254.

Bren L. Frances Oldham Kelsey: FDA medical reviewer leaves her mark on history. *FDA Consumer*. 2001; 35(2) [accessed July 31, 2005]. Available from www .fda.gov/fdac/20/.

Bresalier RS, Sandler RS, Quan H, Bolognese JA, Oxenius B, Horgan K, Lines C, Riddell R, Morton D, Lanas A, Konstam MA, Baron JA, the Adenomatous Polyp Prevention on Vioxx (APPROVe) Trial Investigators. Cardiovascular events associated with rofecoxib in a colorectal adenoma chemoprevention trial. *N Engl J Med*. 2005; 352:1092–1102.

Brown JG. *OIG Report: Experience of Health Maintenance Organizations with Pharmacy Benefits Management Companies*. Washington, DC: Office of Inspector General, Department of Health and Human Services; 1997.

Cabri W, Di Fabio R. *From Bench to Market: The Evolution of Chemical Synthesis*. New York: Oxford University Press; 2000.

Calabresi P, and Chabner B. Chemotherapy of neoplastic disease. In: Hardman JG, Limbied LE, Molinoff BB, Ruddon RW, Gilman AG, eds. *Goodman and Gilman's The Pharmacological Basis of Therapeutics*. New York: McGraw-Hill; 1996 (CD-ROM).

Carpenter G, Cohen S. Epidermal growth factor. *J Biol Chem*. 1990; 265(14):7709–7712.

CASCADE Collaboration (Porter K, Babiker A, Bhaskaran K, Darbyshire J, Pezzotti P, Porter K, Walker AS). Determinants of survival following HIV-1 seroconversion after the introduction of HAART. *Lancet*. 2003; 362:1267–1274.

CBSnews.com. Gov't OKs Vioxx, Celebrex, Bextra. CBS. 2005 [accessed March 25, 2005]. Available from www.cbsnews.com/stories/02/05/heaHG/paintable 671915.shtml.

Center for Drug Evaluation and Research (CDER). Guidelines for the clinical evaluation of anti-inflammatory and antirheumatic drugs [pdf]. FDA, April 1977, revised 1988 [accessed July 28, 2003]. Available from http://www.fda.gov/cder/guidance/old048fn.pdf.

———. Guideline for the clinical evaluation of analgesic drugs [pdf]. FDA, December 1992 [accessed July 28, 2003]. Available from http://www.fda.gov/cder/guidance/old041fn.pdf.

———. New drug evaluation guidance document: refusal to file [pdf]. FDA 1993 [cited November 28, 2003]. Available from http://www.fda.gov/cder/guidance/rtf.pdf.

———. Summary basis for approval for Copaxone (copolymer-1). Bethesda, MD: Food and Drug Administration; 1996. (1996a)

———. In: Caset Associates Ltd., transcribers. *Transcript: Proceedings of the Peripheral and Central Nervous System Drugs Advisory Committee Meeting Number 44*. Gaithersburg, MD: FDA; 1996. (1996b)

———. Guidance for industry: M3 nonclinical safety studies for the conduct of human clinical trials for pharmaceuticals [Internet]. FDA. 1997 [accessed April 11, 2003]. Available from http://www.fda.gov/cder/guidance/1855fnl.pdf.

———. Arthritis Advisory Committee, public hearing: NSAID COX-2 safety issues: FDA. 1998. (1998a)

———. Guidelines for industry: providing clinical evidence of effectiveness for human drugs and biological products [pdf]. FDA. 1998 [accessed September 16, 2003]. Available from http://www.fda.gov/cder/guidance/1397fnl.pdf. (1998b)

———. Slides: Transcript: Arthritis Drugs Advisory Committee meeting. December 1, 1998, Celebrex. FDA. 1998. (1998c)

———. Summary basis for approval for Celebrex (Celecoxib) application No. 20-998. Bethesda, MD: Food and Drug Administration, Center for Drug Evaluation and Research; 1998. (1998d)

———. Transcript: Arthritis Drugs Advisory Committee meeting: open public hearing Washington, DC: US Government Printing Office; 1998. (1998e)

———. Celebrex: Draft Label. Washington, DC: FDA. 1999. (1999a)

———. From test tube to patient: improving health through drugs [pdf]. Food and Drug Administration, September 1999 [accessed July 14, 2003]. Available from http://www.fda.gov/cder/about/whatwedo/testtube.pdf. (1999b)

———. Guidance for industry: qualifying for exclusivity (505A) under the Federal

Food, Drug, and Cosmetic Act. FDA. 1999 [accessed July 31, 2005]. Available from http://www.fda.gov/cder/guidance/289/fnl.pdf. (1999c)

————. Summary basis for approval: Vioxx (Rofecoxib) Tablets; Company: Merck Research Laboratories; Application No.: 021042 & 021052. Washington, DC: FDA; 1999. (1999d)

————. NDA 020998 Supplement 010 Approval Letter/Label. Washington, DC: FDA; 2001. (2001a)

————. Reviewer Diagram Project [Web site html]. FDA. March 8, 2001 [accessed December 15, 2003]. Available from http://www.fda.gov/cder/reviewer/. (2001b)

————. Draft guidance for industry and reviewers: estimating the safe starting dose in clinical trials for therapeutics in adult healthy volunteers. In CBER, eds. Docket No. 02D-0492. Bethesda, MD: FDA; 2002. (2002a)

————. Summary basis for approval: sNDA9 Celebrex (Celecoxib) Capsules; Company: G.D. Searle L.L.C.; Application No.: 20-998/S9; Approval Date: June 7, 2002. Washington, DC: FDA; 2002. (2002b)

————. Career opportunities at CDER: you can make a difference [Web page]. FDA. 2003 [accessed December 8, 2003]. Available from http://www.fda.gov/cder/career/default.htm#Salary. (2003a)

————. The Patent Medicine Menace [Web page]. CDER. 2003 [accessed July 18, 2003]. Available from http://www.fda.gov/cder/about/history/Gallery/galleryintro.htm. (2003b)

————. Pharmaceutical cGMPs for the 21st century—a risk based approach: second progress report and implementation plan. Washington, DC: FDA; 2003. (2003c)

————. Approval Statistics: Cancer. FDA. 2004 [accessed August 4, 2005]. Available from http://www.accessdata.fda.gov/scripts/cder/onctools/statistics.cfm#NME. (2004a)

————. Postmarketing study commitments. FDA. 2004 [accessed March 28, 2004]. Available from http://www.accessdata.fda.gov/scripts/cder/pmc/index.cfm. (2004b)

————. Transcript: Joint Meeting of the Arthritis Advisory Committee with Drug Safety Advisory Committee Day 3, February 18, 2005.

CDER, CBER. Guidance for industry: content and format of investigational new drug application for Phase 1 studies of drugs, including well-characterized, therapeutic biotechnology-derived products. FDA. 1995 [accessed April 22, 2003]. Available from http://www.fda.gov/cder/guidance/phase1.pdf.

————. Guidance for industry: providing regulatory submissions in electronic format—general considerations [pdf]. U.S. Department of Health and Human Services/FDA. 1999 [accessed November 4, 2003]. Available from http://www.fda.gov/cder/guidance/2867fnl.pdf.

Centers for Disease Control (CDC). In: National Center for Health Statistics, ed. 2000. *Leading Causes of Death 1900–1998* [accessed July 31, 2005]. Available from http://www.cdc.gov/nchs/data/dvs/lead1900_98.pdf.

————. *National Vital Statistics Report*. Hyattsville, MD: U.S. Department of Health and Human Services; 2004.

Center for Medicare and Medicaid Services (CMS). Medicaid drug rebate program, May 30, 2003 [accessed February 20, 2004]. Available from http://www.cms .hhs.gov/medicaid/drugs/drughmpg.asp and http://www.cms.hhs.gov/medicaid/ drugs/drug12.asp.

Center for Science in the Public Interest (CSPI). Conflicts of interest on COX-2 panel. CSPINET. 2005 [accessed March 24, 2005]. Available from http://cspinet .org/integrity/press/200502251.html.

Chan A-W, Hrobjartsson A, Haahr MT, Gotzsche PC, Altman DG. Empirical evidence for selective reporting of outcomes in randomized trials: comparison of protocols to published articles. *JAMA*. 2004; 291(20):2457–2465.

Chan FKL, Hung LCT, Suen BY, Wu JCY, Lee KC, Leung VKS, Hui AJ, To KF, Leung WK, Wong VWS, Chung SCS, Sung JJY. Celecoxib versus diclofenac and omeprazole in reducing the risk of recurrent ulcer bleeding in patients with arthritis. *N Engl J Med*. 2002; 347(26):2104–2110.

Choudhry NK, Detsky AS. A perspective on US drug reimportation. *JAMA*. 2005; 293(3):358–362.

Ciardiello F, Tortora G. A novel approach in the treatment of cancer: targeting the epidermal growth factor receptor. *Clin Cancer Res*. 2001; 7(10):2958–2970.

Clinton P. Getting ready for risk-based GMPs. *Biopharm Int*. 2003; 16(4):26.

Code of Federal Regulations (CFR). Title 21, Chapter I, Part 11: Electronic Records; Electronic Signatures. Washington, DC: U.S. Government Printing Office; 2005. (2005a)

————. Title 21, Chapter I, Part 14: Public hearing before a public advisory committee. Washington, DC: Government Printing Office; 2005. (2005b)

————. Title 21, Chapter I, Part 16: Regulatory hearings before the Food and Drug Administration. Washington, DC: U.S. Government Printing Office; 2005. (2005c)

————. Title 21, Chapter I, Part 56: Institutional Review Board. Washington, DC: U.S. Government Printing Office; 2005. (2005d)

————. Title 21, Chapter I, Part 202, Prescription drug advertising. Washington, DC: U.S. Government Printing Office; 2005. (2005e)

————. Title 21, Chapter I, Part 203, Subpart D: Samples. Washington, DC: U.S. Government Printing Office; 2005. (2005f)

————. Title 21, Chapter I, Part 211: Current good manufacturing practice for finished pharmaceuticals. Washington, DC: U.S. Government Printing Office; 2005. (2005g)

————. Title 21, Chapter I, Part 312.22: IND General principles of the IND submission. Washington, DC: U.S. Government Printing Office; 2005. (2005h)

————. Title 21, Chapter I, Part 312.23: IND content and format. Washington, DC: U.S. Government Printing Office; 2005. (2005i)

————. Title 21, Chapter I, Part 312.47: Meetings [pdf]. Washington, DC: U.S. Government Printing Office; 2005. (2005j)

———. Title 21, Chapter I, Part 312.70: Disqualification of a clinical investigator. Washington, DC: U.S. Government Printing Office; 2005. (2005k)

———. Title 21, Chapter I, Part 314.50: Content and format of an application. Washington, DC: U.S. Government Printing Office; 2005. (2005l)

———. Title 21, Chapter I, Part 314.80: Postmarketing reporting of adverse drug experiences. Washington, DC: U.S. Government Printing Office; 2005. (2005m)

———. Title 21, Chapter I, Part 314.101: Filing an application and receiving an abbreviated new drug application. Washington, DC: U.S. Government Printing Office; 2005. (2005n)

———. Orphan Drug Act. Public Law No. 97-414, 96 Stat 2049. Title 21 Chapter I, Part 316. Washington, DC: U.S. Government Printing Office. (2005o)

———. Title 45, Subtitle A: Part 46: Protection of Human Subjects, Subpart C— Additional protections pertaining to biomedical and behavioral research involving prisoners as subjects. Washington, DC: U.S. Government Printing Office; 2005. (2005o)

Comi G, Filippi M, Wolinsky JS. European/Canadian multicenter, double-blind, randomized, placebo-controlled study of the effects of glatiramer acetate on magnetic resonance imaging—measured disease activity and burden in patients with relapsing multiple sclerosis. European/Canadian Glatiramer Acetate Study Group. *Ann Neurol.* 2001; 49(3):290–297.

Consumer Alert. FDA's pediatric rule challenged. Consumer Alert. 1999 [accessed December 11, 2003]. Available from http://www.consumeralert.org/press/pedpr.html.

Contrera JF, Jacobs AC, DeGeorge JJ. Carcinogenicity testing and the evaluation of regulatory requirements for pharmaceuticals*1, *2. *Regul Toxicol Pharmacol.* 1997; 25(2):130–145.

Dannenberg AJ, Lippman SM, Mann JR, Subbaramaiah K, DuBois RN. Cyclooxygenase-2 and epidermal growth factor receptor: pharmacologic targets for chemoprevention. *J Clin Oncol.* 2005; 23(2):254–266.

Danzon PM, Furukawa MF. Prices and availability of pharmaceuticals: evidence from nine countries. *Health Affairs. Suppl. Web Exclusive.* 2003; W3:521–536.

DeGeorge J. Challenges in application of new approaches to carcinogenicity testing for pharmaceuticals. *Toxicol Lett.* 1998; 102–103:565–568.

DeVol RC, Wong P, Bedrossian A, Wallace L, Murphy KJ, Koeppr D. *Biopharmaceutical Industries Contribution to State and U.S. Economics.* Santa Monica, CA: Milken Institute; 2004.

DiMasi JA. Risks in new drug development: approval success rates for investigational drugs. *Clin Pharmacol Ther.* 2001; 69:297–307.

DiMasi JA, Hansen RW, Grabowski HG. The price of innovation: new estimates of drug development costs. *J Health Econ.* 2003; 22(2):151–185.

Dove A. Promising drug is victim of bad business. *Nat Med.* 2002; 8:199.

Drazen JM. COX-2 inhibitors—a lesson in unexpected problems. *N Engl J Med.* 2005; 352:1131–1132.

Druker BJ. STI571 (Gleevec(TM)) as a paradigm for cancer therapy. *Trends Mole Med.* 2002; 8(4):S14–S18.

Eastern Research Group (ERG). Profile of the prescription drug wholesaling industry. Lexington, MA: Economics Staff, Office of Policy, Planning and Legislation, FDA.

Ebert LB. Increasingly aggressive efforts at patent enforcement. *Intellectual Property Today.* June 2000: 22.

Elgie RG. Regulating prices of patented pharmaceuticals in Canada: the patented medicine prices review board. *Food, Drug, Cosmetic Med Device Law Digest.* 1996; 13(2):80–84.

Emanuel EJ, Wendler D, Grady C. What makes clinical research ethical? *JAMA.* 2000; 283(20):2701–2711.

Evenhaim A. Taking e-health relationship management into the next millennium. *Med Marketing Media.* February 26, 2001:10.

FDA. FDA public health advisory: FDA announces important changes and additional warnings for Cox-2 selective and non-selective non-steroidal anti-inflammatory drugs (NSAIDS). Washington, DC: FDA; 2005. Available from http://www.fda.gov/cder/drug/advisory/COX2.htm.

Fisher RA. Statistics. In: Heath AE, ed. *Scientific Thought in the Twentieth Century.* London: Watta; 1951; 31–55.

Fitzgerald GA. Coxibs and cardiovascular disease. *N Engl J Med.* 2004.

Fitzgerald GA, Patrino C. The coxibs, selective inhibitors of cyclooxygenase-2. *N Engl J Med.* 2001; 345(6):433–442.

Food and Drug Administration Modernization Act of 1997. Public Law 105–115. November 20, 1997.

Freedman B. Equipoise and the ethics of clinical research. *N Engl J Med.* 1987; 317(3):141–145.

Fries JF. NSAID gastropathy: the second most deadly rheumatic disease? Epidemiology and risk appraisal. *J Rheumatol.* 1991; 28(suppl):6–10.

From pennies to profits: Interview; Eli Hurvitz. *Institutional Investor.* February 1998: 21–23.

Gierse JK, Hauser SD, Creely DP, Koboldt C, Rangwala SH, Isakson PC, Seibert K. Expression and selective inhibition of the constitutive and inducible forms of human cyclo-oxygenase. *Biochem J.* 1995; 305:479–484.

Gittins J. Quantitative methods in the planning of pharmaceutical research. *Drug Inf J.* 1996; 30:479–487.

Golde ER. Advising under the influence?: conflicts of interest among FDA advisory committee members. *Food Drug Law J.* 2002; 57:293–322.

Goldkind L. Medical officer's gastroenterology advisory committee briefing document NDA 20,998 supplement #9. Washington, DC: FDA; 2001.

Goldsworthy P, McFarlain AC. Howard Florey, Alexander Fleming and the fairy tale of penicillin. *Med J Aust.* 2002; 176:178–180.

Goodman B. Do drug company promotions influence physician behavior? *West J Med.* 2001; 174:232–233.

Goodman L, Wintrobe MM, Damesheck W, Goodman MJ, Gilman A, McLennan MT. Nitrogen mustard therapy: use of methyl-bis(beta-chloroethyl)amine hydrochloride and tris(beta-chloroethyl)amine hydrochloride for Hodgkin's disease, lymphosarcoma, leukemia and certain allied and miscellaneous disorders. *JAMA.* 1946; 132:126–132.

Grabowski HG. Pharmaceuticals: politics, policy and availability: patents and new product development in the pharmaceutical and biotechnology industries. *Georgetown Public Policy Rev.* 2003; 8(2):7–22.

Grabowski HG, Vernon JM. A new look at the returns and risks to pharmaceutical R&D. *Manage Sci.* 1990; 38(7):804–821.

Graham DJ, Campen D, Hui R, Spence M, Cheetham C, Levy G, Shoor S, Ray WA. Risk of acute myocardial infarction and sudden cardiac death in patients treated with cyclo-oxygenase 2 selective and non-selective non-steroidal anti-inflammatory drugs: nested case-control study. *Lancet.* 2005; 365(9458): 475–481.

Groopman J. Superaspirin. *The New Yorker.* June 15, 1998: 32–35.

———. Celebra: new, safe drug for arthritic pain. *Saturday Evening Post.* September/October 1998; 270:60.

Haffner M. Memorandum: Office of Orphan Products Development (OOPD) analysis of exclusivity issues raised in the Serono BLA for Rebif. Washington, DC: FDA; 2002.

Harris JO, Frank JA, Patronas N, McFarlin DE, McFarland HF. Serial gadolinium-enhanced magnetic resonance imaging scans in patients with early, relapsing-remitting multiple sclerosis: implications for clinical trials and natural history. *Ann Neurol.* 1991; 29(5):548–555.

Harris R. *The Real Voice.* New York: Macmillan Company; 1964.

Hawkey CJ. Cyclooxygenase inhibition: between the devil and the deep blue sea. *Gut.* 2002; 50(90003):25–30.

HDMA. *Healthcare Product Distribution: A Primer.* Reston, VA: HDMA; 2003.

Health Research Extension Act of 1985. Public Law 99-158. November 20, 1985.

Herper M. Five ways to fix the FDA. Forbes.com. January 12, 2005 [accessed August 1, 2005]. Available from http://www.forbes.com/2005/01/12/cx_mh_0112fdaintro.htm

Hiam A. *Marketing for Dummies.* New York: Hungry Minds; 1997.

Hilts PJ. Seeking limits to a drug monopoly. *New York Times.* May 14, 1992; Page 1, Column 3.

———. *Protecting America's Health: The FDA, Business and One Hundred Years of Regulation.* New York: Alfred A. Knopf; 2003.

Hoffmann C, Kamps BS, eds. *HIV Medicine 2003.* Paris: Flying Publisher; 2003.

Honig S. Outline for performing NDA review. CDER. 2003 [accessed December 14, 2003]. Available from http://www.fda.gov/cder/reviewer/honig.pdf.

Hornblum AM. They were cheap and available: prisoners as research subjects in twentieth century America. *Br Med J.* 1997; 315:1437–1441.

Hróbjartsson A, Gøtzsche PC. Is the placebo powerless?—an analysis of clinical trials comparing placebo with no treatment. *N Engl J Med.* 2001; 344(21): 1594–1602.

IMS. IMS reports 11.5 percent dollar growth in 2003 U.S. prescriptions sales: growth is sustained by new products despite a difficult year. *Business Wire.* 2004.

Insel PA. Analgesic-antipyretics and antiinflammatory agents and drugs employed in the treatment of gout. In: Hardman JG, Limbird LE, Molinoff PB, Ruddon RW, and Gilman AG, eds. *Goodman and Gilman's The Pharmacological Basis of Therapeutics.* New York: McGraw-Hill; 1996 (CD-ROM).

Institute of Medicine (IOM). *The Food and Drug Administration Advisory Committees.* Washington, DC: National Academies Press; 1992.

International Conference on Harmonization (ICH). ICH harmonized tripartite guideline: clinical safety data management; definitions and standards for expedited reporting: international conference on harmonization of technical requirements for registration of pharmaceuticals for human use. Geneva, Switzerland: ICH; 1994.

———. S1A: Guideline on the need for carcinogenicity studies of pharmaceuticals: international conference on harmonization of technical requirements for registration of pharmaceuticals for human use. 1995 [accessed 2004]. Available from http://www.ich.org/MediaServer.jser?@_ID=489&@_MODE=GLB.

———. E6: Good clinical practice: consolidated guideline: international conference on harmonization of technical requirements for registration of pharmaceuticals for human use. 1996 [accessed August 1, 2005]. Available from http://www.ich.org/MediaServer.jser?@_ID=482&@_MODE=GLB.

———. A brief history of the ICH: international conference on communications. 1997 [accessed April 11, 2003]. Available from http://www.ich.org/ich8.html#History. (1997a)

———. M3: Maintenance of the ICH guidelines on non-clinical safety studies for the conduct of human clinical trials for pharmaceuticals: international conference on harmonization of technical requirements for registration of pharmaceuticals for human use. 1997 [accessed May 27, 2003]. Available from http://www.ich.org/pdfICH/m3mstep4.pdf.

———. Safety pharmacology studies for human pharmaceuticals: international conference on harmonization of technical requirements for registration of pharmaceuticals for human use. 2000 [accessed May 27, 2003]. Available from http://www.ich.org/pdfICH/S7step4.pdf.

———. MedDRA® term selection: points to consider [pdf]: international conference on harmonization, July 18, 2003 [accessed December 16, 2003].

————. ICH E2E: Pharmacovigilance planning (PvP). 2005.

Jackson K, Young D, Pant S. Drug-excipient interactions and their effect on absorption. *Pharm Sci Technol Today.* 2000; 3(10):336–345.

Janssen WF. The story of the laws behind the labels: Part I, 1906 Food and Drugs Act [Web page]. FDA Consumer, June 1981 [accessed July 20, 2002]. Available from http://vm.cfsan.fda.gov/~lrd/history1.html. (1981a)

————. The story of the laws behind the labels: Part II, 1938—The Federal Food, Drug, and Cosmetic Act. FDA Consumer, 1981 [accessed July 20, 2003]. Available from http://vm.cfsan.fda.gov/~lrd/historla.html. (1981b)

————. The story of the laws behind the labels: Part III, 1962 Drug Amendments. FDA Consumer, 1981 [accessed July 20, 2003]. Available from http://vm.cfsan.fda .gov/~lrd/historlb.html. (1981c)

Johannes L. Smart marketing helps Biogen match Schering. *Wall Street Journal,* December 19, 1996; 10.

Johns Hopkins University. Adenomatous Polyposis of the Colon [accessed July 30, 2001]. Available from http://www.ncbi.nlm.nih.gov/entrez/dispomim.cgi? id=175100.

Johnson KP, Brooks BR, Cohen JA, Ford CC, Goldstein J, Lisak RP, Myers LW, Panitch HS, Rose JW, Schiffer RB. The Copolymer 1 Multiple Sclerosis Study Group. Copolymer 1 reduces relapse rate and improves disability in relapsing-remitting multiple sclerosis: results of a phase III multicenter, double-blind placebo-controlled trial. *Neurology.* 1995; 45(7):1268–1276.

Johnson KP, Brooks BR, Cohen JA, Ford CC, Goldstein J, Lisak RP, Myers LW, Panitch HS, Rose JW, Schiffer RB, Vollmer T, Weiner LP, Wollinsky JS. Copolymer 1 Multiple Sclerosis Study Group. Extended use of glatiramer acetate (Copaxone) is well tolerated and maintains its clinical effect on multiple sclerosis relapse rate and degree of disability. *Neurology.* 1998; 50(3): 701–708.

Jonsen AR. *The Birth of Bioethics.* New York: Oxford University Press; 1998.

————. *A Short History of Medical Ethics.* New York: Oxford University Press; 2000.

Jouzeau JY, Terlain B, Abid A, Nedelac E, Netler P. Cyclo-oxygenase isoenzymes. How recent findings affect thinking about nonsteroidal anti-inflammatory drugs. *Drugs.* 1997; 53(4):563–582.

Kaiser Family Foundation. *Prescription Drug Trends: A Chartbook Update.* Menlo Park, CA: The Kaiser Family Foundation; 2001.

Kalb PE, Dunlop KO, McEnroe DC, Stein SD. Direct-to-consumer marketing: the Food and Drug Administration is not alone. *Food Drug Law J.* 2003; 58:25–33.

Kaptchuk TJ. Intentional ignorance: a history of blind assessment and placebo controls in medicine. *Bull Hist Med.* 1998; 72(3):389–433.

————. The double-blind, randomized, placebo-controlled trial: gold standard or golden calf. *J Clin Epidemiol.* 2001; 54:541–549.

Katz D, Caplan AL. All gifts large and small: toward an understanding of the ethics of pharmaceutical gift-giving. *Am J Bioethics.* 2003; 3(3):39–46.

Katz D, Deshpande M. An Rx for the modification of the Medicare Prescription Drug, Improvement, and Modernization Act of 2003: toward a reform with results. *Ann Health Law.* 2005; 14 (Winter):183–202.

Keith AB, Arnon R, Teitelbaum D, et al. The effect of Cop-1, a synthetic polypeptide, on chronic relapsing experimental allergic encephalomyelitis in guinea pigs. *J Neurol Sci.* 1979; 42:267–274.

Kessler DA, Rose JL, Temple RJ, Schapiro R, Griffin JP. Therapeutic-class wars—drug promotion in a competitive marketplace. *N Engl J Med.* 1994; 331(20): 1350–1353.

Kieffer RG. Procedures: improving their quality. *Pharm Technol.* January 1, 2003: 64–71.

Kitamura Y. Increased expression of cyclooxygenases and peroxisome proliferator-activated receptor in Alzheimer's Disease Brains. *Biochem Biophys Res Commun.* 1999; 254(3):582–586.

Kreling DH, Mott DA, et al. Prescription drug trends: a chartbook update. The Kaiser Family Foundation. 2001; Publication Number: 3112. Available from http://www.kff.org/rxdrugs/3112-index.cfm.

Kukich S. NDA review process overview [pdf, online]. FDA, Division of Anti-Viral Drug Products. 2003 [accessed December 2, 2003]. Available from http://www.fda.gov/cder/reviewer/kukich.pdf.

Kumar P, Zaugg A-M. IMS review: steady but not stellar. *Medical Marketing & Media.* 2003; 50–63.

Landau R, Achilladelis B, Scarabine A, eds. *Pharmaceutical Innovation: Revolutionizing Human Health.* Philadelphia: Chemical Heritage Foundation; 1999.

Langreth R. Drug makers zero in on new treatments that may slow the progress of arthritis. *Wall Street Journal.* May 21, 1996: 1.

La Porte J-R, Dotres C, Diogène E, Peña JP, Reggi V, Márque M. Cuba: improving use of medicines. *Lancet.* 1997; 349:SIII4.

Lawrence S. Measure marketing. *Acumen J Life Sci.* 2003; 1(4):3–5.

Lederer SE. *Subjected to Science: Human Experimentation in America Before the Second World War.* Baltimore: Johns Hopkins University Press; 1995.

Lemonick MD. Arthritis under arrest: new treatments may finally succeed in putting one of the worst forms of this painful illness on ice. *Time Magazine.* September 28, 1998 [accessed online January 31, 2004]. Available from http://www.time.com/time/archive/preview/0,10987,989175,00.html

Lepore PD. FDA's good laboratory practice regulations and computerized data acquisition systems. *Chemometrics Intell Lab Syst.* 1992; 17(3):265–282.

Levesque LE, Brophy JM, Zhang B. The risk for myocardial infarction with cyclooxygenase-2 inhibitors: a population study of elderly adults. *Ann Intern Med.* 2005; 142:481–489.

Levine RJ. *Ethics and Regulation of Clinical Research.* Second ed. New Haven, CT: Yale University Press; 1988. Original edition, 1986.

Lichtenberg FR, Philipson TJ. *The Dual Effects Of Intellectual Property Regulations:*

Within and Between Patent Competition in the US Pharmaceutical Industry. Cambridge, Mass.: National Bureau of Economic Research; 2002.

Liebman M. Head-to-head marketing . . . may the best promoted drug win. *Medical Marketing & Media.* November 1, 2000: 94–98.

Lipsky PE, Isakson PC. Outcome of specific COX-2 inhibition in rheumatoid arthritis. *J Rheumatol.* 1997; 24(suppl 49):9–14.

Livingston RB. *Progress Report on Survey of Moral and Ethical Aspects of Clinical Investigation.* Bethesda, MD: NIH; 1964.

Loehrer P, Einhorn LH. Cisplatin: diagnosis and treatment: drugs five years later. *Ann Intern Med.* 1984; 100:704–713.

Ma J, Stafford R, Cockburn IM, Finkelstein SN. A statistical analysis of the magnitude and composition of drug promotion in the United States in 1998. *Clin Ther.* 2003; 25:1503–1517.

Maggos C. Tysabri war plans. *BioCentury.* December 6, 2004: 1–5.

Masferrer JR, Seibert K, Zweifel B, Needleman P. Endogenous glucocorticoids regulate an inducible cyclooxygenase enzyme. *Proc Natl Acad Sci USA.* 1992; 89:3917–3921.

Mathews AW. Worrisome ailment in medicine: misleading journal articles. *Wall Street Journal.* May 10, 2005: A1.

Mathews AW, Hensley S. Merck may return Vioxx to market. *Wall Street Journal.* February 18, 2005: 3. (2005a)

———. FDA stiffens painkiller warnings, pushes Pfizer to suspend Bextra. *Wall Street Journal.* April 8, 2005: A1. (2005b)

Mathieu M. *New Drug Development: A Regulatory Overview.* Third ed. Waltham, MA: Parexel; 1994.

McCellan M. *The Food and Drug Administration's Strategic Action Plan Protecting and Advancing America's Health: Responding to New Challenges and Opportunities.* Washington, DC: FDA; 2003.

McClellan K, Jarvis B. Desloratadine. *Drugs.* 2001; 61(6):789–796.

Meadows M. MedWatch: managing risks at the FDA (September–October 2003). FDA. 2003 [accessed 2005]. Available from http://www.fda.gov/fdac/features/2003/503_risk.html.

Medicare Prescription Drug, Improvement, and Modernization Act of 2003. Public Law 108-173. January 7, 2003.

Mehrotra R, et al. Global multiple sclerosis survey: steady growth ahead. London: SG Cowen; 2003.

Merck & Co. claim victory in cox-2 patent dispute. *Pharma Marketletter.* March 11, 2002.

MMWR. Ten great public health achievements—United States, 1900–1999. *Morb Mortal Wkly Rep.* 1999; 48(12):241–243.

Morning Briefing. *St. Louis Post-Dispatch.* February 24, 1999: C2.

Mukherjee D, Nissen SE, Topol EJ. Risk of cardiovascular events associated with selective COX-2 inhibitors. *JAMA.* 2001; 286(8):954–959.

Myers CD, Robinson ME, Riley JL III, Sheffield D. Sex, gender, and blood pressure: contributions to experimental pain report. *Psychosom Med.* 2001; 63(4): 375, 545.

Nader R, Love J. Federally funded pharmaceutical inventions: testimony before the Special Committee on Aging of the United States Senate, 1993 [accessed August 1, 2005]. Available from http://www.cptech.org/pharm/pryor .html.

National Association of Boards of Pharmacy (NABP). [Web site] 2004 [accessed February 15, 2004]. Available from http://www.nabp.net/.

National Association of Chain Drug Stores (NACDS). Pharmacy sales. NACDS, 2003 [accessed February 13, 2004]. Available from http://www.nacds.org/ user-assets/PDF_files/Pharmacy_Sales.pdf.

National Commission for the Protection of Human Subjects of Biomedical and Behavioral Research. The Belmont Report: ethical principles and guidelines for the protection of human subjects of research. Department of Health, Education, and Welfare. 1979 [accessed June 27, 2003]. Available from http://ohrp .osophs.dhhs.gov/humansubjects/guidance/belmont.htm.

National Institute for Health Care Management Research and Educational Foundation (NIHCMF). Prescription drugs and mass media advertising [pdf]. National Institute for Health Care Management Research and Educational Foundation. 2000 [accessed February 5, 2004]. Available from http://www.nihcm.org /DTC brief.pdf

National Multiple Sclerosis Society (NMSS). About MS. National Multiple Sclerosis Society. 2003 [accessed March 31, 2003]. Available from http://www .nationalmssociety.org/about%20ms.asp.

National Organization for Rare Disorders (NORD). Familial adenomatous polyposis [Web page]. 2001 [accessed March 20, 2004]. Available from http://hw.health dialog.com/kbase/nord/nord142.htm#nord142-standard-therapies.

National Research Council (NRC). *Drug Efficacy Report: Final Report to the Commissioner of Food and Drugs, Food and Drug Administration.* Washington, DC: National Academy of Sciences; 1969.

NEJM. Looking back on the millennium in medicine. *N Engl J Med.* 2000; 342 (1):42–49.

Nuremberg Military Tribunal. Nuremberg Code. Washington, DC: U.S. Government Printing Office; 1949.

Nussmeier NA, Whelton AA, Brown MT, Langford RM, Hoeft A, Parlow JL, Boyce SW, Verburg KM. Complications of the COX-2 inhibitors parecoxib and valdecoxib after cardiac surgery. *N Engl J Med.* 2005; 352:1081–1091.

Office of Technology Assessment (OTA), U.S. Congress. *Pharmaceutical R&D: Costs, Risks and Rewards.* Washington, DC: U.S. Government Printing Office; 1993.

Orlowski JP, Wateska L. The effects of pharmaceutical firm enticements on physician prescribing patterns. There's no such thing as a free lunch. *Chest.* 1992; 102(1):270–273.

Ott E, Nussmeier NA, Duke PC, Feneck RO, Alston RP, Snabes MC, Hubbard RC, Hsu PH, Saidman LJ, Mangano DT. Efficacy and safety of the cyclooxygenase 2 inhibitors parecoxib and valdecoxib in patients undergoing coronary artery bypass surgery. *J Thorac Cardiovasc Surg.* 2003; 125(6):1481–1492.

Pain Management. *Medical & Healthcare Marketplace Guide.* Online database: RDS-ACC-NO-1349409. January 1997.

Papic RJ. Origins of cancer therapy, a medical review. *Yale J Biol Med.* 2001; 74:391–398.

Parkins DA, Lashmar UT. The formulation of biopharmaceutical products. *Pharm Sci Technol Today.* 2000; 3(4):129–137.

Penning TD, Talley JT, Bertenshaw SR, Carter JS, Collins PW, Doctor S, Graneto MJ, Lee LF, Malecha JW, Miyashiro JM, Rogers RS, Rogier DJ, Yu SS, Anderson GD, Burton EG, Cogburn JN, Gregory SA, Koboldt CM, Perkins W, Siebert K, Veenhuizen AW, Zhang YY, Isakson PC. Synthesis and biological evaluation of the 1,5-diarylpyrazole class of cyclooxygenase-2 inhibitors: identification of 4-[5-(4-methylphenyl)-3-(trifluoromethyl)-1H-pyrazol-1-y] benzenesulfonamide (SC-58635, Celecoxib). *J Med Chem* 1997. 40:1347–1365.

Pfizer. Advisory Committee Briefing Document: Celecoxib and Valdecoxib Cardiovascular Safety. New York: Pfizer; 2005.

Pfizer/Pharmacia. Celebrex Prescribing Information. New York: Pfizer/Pharmacia; 2002.

Pham A. New arthritis drug sells briskly; insurers fear a crippled bottom line [wire article]. Knight-Ridder; 1999 [accessed March 4, 2004].

Pharmaceutical Research and Manufacturers Association (PhRMA). Pharmaceutical Industry Profile, 2004. Pharmaceutical Research and Manufacturers Association. 2004 [accessed August 3, 2005]. Available from http://www.phrma.org/publications/publications//2004-03-31.937.pdf.

Pinchasi I. Interview with Bernice Schacter, June 30, 2003.

Placentra M. New prescriptions plummet graph in Vrazof Fitzgerald, Ginsberg painkillers . . . health. *Philadelphia Inquirer.* February 15, 2005: A01.

Pollack A. F.D.A. approves cancer drug from Genentech. *New York Times*, February 27, 2004: 1.

Porter R. *The Greatest Benefit to Mankind.* New York: W.W. Norton & Company; 1997.

Prasit P, Wang Z, Brideau C, Chan CC, Charleson S, Cromlish W, Ethier D, Evans JF, Ford-Hutchinson AW, Gauthier JY, Gordon R, Guay J, Gesser M, Kargman S, Kennedy B, Leblanc Y, Leger S, Mancini J, O'Neill GP, Ouellet M, Percival MD, Perrier H, Riendeau D, Roger I, Zamboni R, Boyce S, Rupniak N, Forrest M, Visco D, Patrick D. The discovery of Rofecoxib [MK 966, VIOXX, 4-(4"-methylsulfonylphenyl)-3-phenyl-2(5H)-furranone], an orally active cyclooxygenase-2 inhibitor. *Bioorg Med Chem Lett.* 1999; 9:1773–1778.

Public Health Security and Bioterrorism Preparedness and Response Act. Public Law 107-188. 2002.

Pusinelli GA. The Orphan Drug Act—what's right with it? *Santa Clara Comput High Technol Law J.* 1999; 15:299–345.

Randall T. Kennedy hearings say no more free lunch—or much else—from drug firms. *JAMA.* 1991; 265:440–442.

Repic O. *Principles of Process Research and Chemical Development in the Pharmaceutical Industry.* New York: John Wiley & Sons; 1998.

Reverby SM. More than fact and fiction: cultural memory and the Tuskegee Syphilis Study. *Hastings Cent Rep.* 2001; 31(5):22–28.

Reynolds T. Pharmaceutical companies adopt new guidelines for marketing. *J Natl Cancer Inst.* 2002; 94(15):1119–2012.

Riendeau D, Percival MD, Brideau C, Charleson S, Dube D, Ethier D, Falgueyret J-P, Friesen RW, Gordon R, Greig G, Guay J, Mancini J, Ouellet M, Wong E, Xu L, Boyce S, Visco D, Girard Y, Prasit P, Zamboni R, Rodger IW, Gresser M, Ford-Hutchinson AW, Young RN, Chan CC. Etoricoxib (MK-0663): preclinical profile and comparison with other agents that selectively inhibit cyclooxygenase-2. *J Pharmacol Exp Ther.* 2001; 296(2):558–566.

Robinson K. GLPs and the importance of standard operating procedures. *BioPharm Int.* 2003; 16(8):38.

Rock A. Pain, pain go away: is Celebrex—the new arthritis drug—all it's cracked up to be? *Money.* 1999; 28(4):196–197.

Rodriguez D. Decisions of pharmaceutical firms for new product development. Working Paper 98-006WP. Cambridge, MA: MIT Industrial Performance Center; 1998.

Rolak L. The history of MS: the basic facts [html]. National Multiple Sclerosis Society. 2003 [accessed March 31, 2003]. Available from http://www.nationalmssociety.org/Brochures-HistoryofMS1.asp.

Rosenberg R. Record profits for Biogen. *Boston Globe.* October 10, 1996: 1.

———. Biogen's Avonex top drug treatment for MS. *Boston Globe.* January 14, 1998: 9.

Rosenthal MB, Berndt ER, Donohue JM, Frank RG, Epstein AM. Promotion of prescription drugs to consumers. *N Engl J Med.* 2002; 346(7):498–505.

Rubinsztein DC. How useful are animal models of human disease? *Semin Cell Dev Biol.* 2003; 14(1):1–2.

Santora M. U.S. is close to eliminating AIDS in infants, officials say. *New York Times.* January 30, 2005: 1.

Scharf S, Mander A, Ugoni A, Vajda F, Christophidis N. A double-blind, placebo-controlled trial of diclofenac/misoprostol in Alzheimer's disease. *Neurology.* 1999; 53(1):197–201.

Schlessinger J, Givol D, Bellot F, Kris R, Ricca GA, Cheadle C, South V, inventors U.S. Patent. 6,217,866 Monoclonal antibodies specific to human epidermal growth factor receptor and therapeutic methods employing same. Rhone-Poulenc Rorer International (Holdings), Inc., assignee; 2001.

Searle and Pfizer announce agreement to co-promote innovative antiarthritis agents. *PR Newswire Chicago.* February 18, 1998.

Searle and Pfizer deepen celecoxib agreement. *Pharmaceutical Business News.* March 25, 1998.

Sevach EM. Organ-specific autoimmunity. In: Paul WS, ed. *Fundamental Immunology.* Philadelphia: Lippincott-Raven; 1999.

Shillingford CA, Vose CW. Effective decision-making: progressing compounds through clinical development. *Drug Discovery Today.* 2001; 6(18):941–946.

Shuster E. Fifty years later: the significance of the Nuremberg Code. *N Engl J Med.* 1997; 337(20):1436–1440.

Siebert K, Zhan Y, Leahy K, Hauser S, Masferrer J, Perkins LL, Isakson P. Pharmacological and biochemical demonstration of the role of cyclooxygenase 2 in inflammation and pain. *Proc Nat Acad Sci USA.* 1994; 91:12013–12017.

Sieker B. Place. In: Kolassa EM, Smith MC, Perkins G, Sieker B, eds. *Pharmaceutical Marketing: Principles, Environment and Practice.* New York: Pharmaceutical Products Press; 2002: sect IV.

Silverstein AM. *A History of Immunology.* San Diego: Academic Press; 1989.

Silverstein FE, Faich G, Goldstein JL, Simon LS, Pincus T, Whelton A, Makuch R, Eisen G, Agrawal NM, Stenson WF, Burr AM, Zhao WW, Kent JD, Lefkowith JB, Verburg KM, Geis GS. Gastrointestinal toxicity with Celecoxib vs nonsteroidal anti-inflammatory drugs for osteoarthritis and rheumatoid arthritis: the CLASS study: a randomized controlled trial. *JAMA.* 2000; 284(10):1247–1255.

Silverstein FE, Graham DY, Senior JR, Davies HW, Struthers BJ, Bittman RM, Geis GS. Misoprostol reduces serious gastrointestinal complications in patients with rheumatoid arthritis receiving nonsteroidal anti-inflammatory drugs: a randomized, double-blind, placebo-controlled trial. *Ann Intern Med.* 1995; 123(4):241–249.

Simon LS, Lanza FL, Lipsky PE, Hubbard RC, Talwalker S, Schwartz BD, Isakson PC, Geis GS. Preliminary study of the safety and efficacy of SC-58635, a novel cyclooxygenase 2 inhibitor: efficacy and safety in two placebo-controlled trials in osteoarthritis and rheumatoid arthritis, and studies of gastrointestinal and platelet effects. *Arthritis Rheum.* 1998; 41(9):1591–1602. (1998a)

Sinicrope FA, Gill S. Role of cyclooxygenase-2 in colorectal cancer. *Cancer Metastasis Rev.* 2004; 23(1–2):63–75.

Smith M. Pharmaceutical Promotion Practices. In: Smith MC, Kolassa EM, Perkins G, Sieker B, eds. *Pharmaceutical Marketing: Principles, Environment, and Practice.* New York: Pharmaceutical Products Press; 2002: chap 15.

Solomon DH, Avorn J. Coxibs, science, and the public trust. *Arch Intern Med.* 2005; 165:158–160.

Solomon S. Diagnosing Eli Hurvitz's headaches. *The Jerusalem Post.* March 3, 1998: 14.

Solomon SD, McMurray JJV, Pfeffer MA, Wittes J, Fowler R, Finn P, Anderson WF, Zauber A, Hawk E, Bertagnolli M, the Adenoma Prevention with celecoxib (APC) Study Investigators. Cardiovascular risk associated with Celecoxib in a clinical trial for colorectal adenoma prevention. *N Engl J Med*. 2005; 352:1071–1080.

SPP. *Country Report: Cuba in Transition*. Ann Arbor, MI: Gerald R. Ford School of Public Policy, University of Michigan; 2004: 674.

Srinivasu MK, Sreenivas Rao D, Om Reddy G. Determination of celecoxib, a COX-2 inhibitor, in pharmaceutical dosage forms by MEKC. *J Pharma Biomed Anal*. 2002; 28(3–4):493–500.

Starr P. *The Social Transformation of American Medicine*. New York: Basic Books; 1982.

Steinbach G, Lynch PM, Phillips RKS, Wallace MH, Hawk E, Gordon GB, Wakabayashi N, Saunders B, Shen Y, Fujimura T, Su L-K, Levin B, Godio L, Patterson S, Rodriguez-Bigas MA, Jester SL, King KL, Schumacher M, Abbruzzese J, DuBois RN, Hittelman WN, Zimmerman S, Sherman JW, Kelloff G. The effect of celecoxib, a cyclooxygenase-2 inhibitor, in familial adenomatous polyposis. *N Engl J Med*. 2000; 342(26):1946–1952.

Stewart WF, Kawas C, Corrad M, Mettler EJ. Risk of Alzheimer's disease and duration of NSAID use. *Neurology*. 1997; 48(3):626–632.

Steyer R. Monsanto readies sales blitz for new drug. *St. Louis Post-Dispatch*. February 13, 1999: Business 32. (1999a)

Swann JP. History of the FDA. U.S. Food and Drug Administration. 1998 [accessed June 15, 2003]. Available from http://www.fda.gov/oc/history/historyoffda/fulltext.html.

Talley J. Interview with Bernice Schacter, June 23, 2003.

Tancer RS. The pharmaceutical industry in Cuba. *Clin Ther*. 1995; 17(4):791–798.

Teitelbaum D, Arnon R, Sela M. Suppression of experimental allergic encephalomyelitis by a synthetic polypeptide. *Eur J Immunol*. 1971; 1:242–248.

———. Copolymer 1: from basic research to clinical applications. *Cell Mol Life Sci*. 1997; 53:24–28.

Teitelbaum D, Gan R, Meshorer A, Hirshfeld T, Arnon R, Sela M. Inventors. Therapeutic copolymer. Yeda Research and Development Co. Rehovot, Israel, assignee. 1974a; 3,849,550.

Teitelbaum D, Webb C, Bree M, Meshorer A, Arnon R, Sela M, Suppression of experimental allergic encephalomyelitis in rhesus monkeys by a synthetic basic copolymer. *Clin Immunol Immunopathol*. 1974b; 3:256–262.

Teva Neurosciences. Copaxone (glatiramer acetate injection). Package Insert. 2002.

Teva Pharmaceuticals Inc. The History of Teva [Web page]. Teva Pharmaceuticals. 2003 [accessed July 6, 2003]. Available from http://www.tevapharm.com/about/history.asp.

Tinsley H. Prescriptions without Borders: America looks to Canada for answers to solve the prescription drug pricing predicament in the U.S., but is importation really the solution? *Hamline J Public Law Policy.* 2004; 25:437–479.

Tirronen E, Salmi T. Process development in the fine chemical industry. *Chem Eng J.* 2003; 91(2–3):103–114.

Topol EJ. Arthritis medicines and cardiovascular events—"house of coxibs." *JAMA.* 2005; 293(3):366–368.

Trost BM. The atom economy—a search for synthetic efficiency. *Science.* 1991; 254:1471–1477.

University of Rochester v. G.D. Searle & Co., Inc. 2003. In LEXIS: United States District Court for the Western District of New York.

Uruguay Round Agreements Act. Public Law 103-465. December 8, 1994.

U.S. Code. Title 18. Crimes and Criminal Procedure, Part I: Crimes; Chapter 11: Bribery, Graft and Conflicts of Interest, Section 208: Acts affecting a personal financial interest, 2001.

U.S. Code. Title 21. Chapter 9. Subchapter VIII. Section 381: Imports and exports, 2005. Available from http://assembler.law.cornell. edu/uscode/21/381/htm.

U.S. Code. Title 35. Part III—Patents and protection of patent rights. Ithaca, NY: Legal Information Institute, Cornell University; 2003.

U.S. Congress. Health and Environment Subcommittee of the House Energy and Commerce Committee. Hearing: FDA User Fees for Prescription Drug Approval. 1992.

U.S. Department of Health and Human Services (DHHS). Establishment of prescription drug user fee rates for fiscal year 2005. Federal Register. Washington, DC: DHHS; 2004.

U.S. Food and Drug Administration (FDA). Nonclinical laboratory studies: proposed regulations for good laboratory practice. Federal Register. Washington, DC: U.S. Government Printing Office; 1976.

———. Good laboratory practice regulations effective June 1979, and amended effective October 1987. Food and Drug Administration, Department of Health and Human Services. 1987 [accessed April 11, 2003]. Available from http://www.access.gpo.gov/nara/cfr/waisidx_02/21cfr58_02.html.

———. Bioresearch monitoring program coordination background: FDA; 1998. (1998a)

———. FDA operations: FDA inspections of clinical investigators. FDA. 1998 [accessed December 19, 2003]. Available from http://www.fda.gov/oc/ohrt/irbs/operations.html. (1998b)

———. Guidance for industry: consumer-directed broadcast advertisements. Rockville, MD: FDA; 1999.

———. FDA guidance on conflict of interest for advisory committee members, consultants and experts. Washington, DC: FDA; 2000. (2000a)

———. Guidance for industry formal meetings with sponsors and applicants for PDUFA Products. Washington, DC: Department of Health and Human Services; 2000. (2000b)

———. Report to Congress: The Prescription Drug Marketing Act. Washington, DC: Department of Health and Human Services; 2001. (2001a)

———. *CDER 2002 Report to the Nation: Improving Public Health Through Human Drugs.* Rockville, MD: Food and Drug Administration; 2002. (2002a)

———. Chapter I—Food and Drug Administration, Department of Health and Human Services, Part 50—Protection of Human Subjects. Washington, DC: U.S. Government Printing Office; April 2002 [accessed July 11, 2002]. Available from http://www.access.gpo.gov/nara/cfr/waisidx_02/21cfr50_02.html. (2002b)

———. Direct to Consumer Promotion: Public Meeting; 2003. (2003a)

———. FDA Form 1571 Cover pages for Investigational New Drug Application. FDA. 2003 [accessed June 19, 2003]. Available from http://forms.psc.gov/forms/FDA/FDA-1571.pdf. (2003b)

———. FY 2003 Performance Report to the President and the Congress. Washington, DC: DHHS; 2003. (2003c)

———. List of all orphan products designations and approvals [MSWord document]. Office of Orphan Products Development. 2003 [accessed July 30, 2005]. Available from http://www.fda.gov/orphan/designat/alldes.rtf. (2003d)

———. PDUFA III five-year plan (July 2003) [online]. Office of Management and Systems. 2003 [accessed December 2, 2003]. Available from http://www.fda.gov/oc/pdufa3/2003plan/default.htm. (2003e)

———. Electronic Orange Book: approved drug products with therapeutic equivalence evaluations. HHS. March 19, 2004 [accessed March 25, 2004]. Available from http://www.fda.gov/cder/ob/default.htm. (2004a)

———. MedWatch: The FDA safety information and adverse event reporting program. 2004 [accessed August 1, 2005]. Available from http://www.fda.gov/medwatch/index.html. (2004b)

———. FDA Talk Paper: 2004 FDA accomplishments. FDA [accessed March 25, 2005]. Available from http://www.fda.gov/bbs/topics/Answers/2005/ANS01346.html. (2005a)

———. Press Office. FDA Improvements in Drug Safety Monitoring. FDA Web site. FDA. 2005. (2005b)

U.S. General Accounting Office (GAO). Prescription drugs: companies typically charge more in the United States than in Canada. Washington, DC: General Accounting Office; 1992.

———. Prescription drugs: companies typically charge more in the United States than in the United Kingdom. Washington, DC: General Accounting Office; 1994.

———. FDA drug approval review time has decreased in recent years. Washington, DC: General Accounting Office; 1995.

———. FDA oversight of direct-to-consumer advertising has limitations. Washington, DC: General Accounting Office; 2002. (2002a)

———. Food and Drug Administration: Effect of user fees on drug approval times, withdrawals, and other agency activities. Washington, DC: General Accounting Office; 2002. (2002b)

U.S. Patent and Trademarks Office. What are patents, trademarks, servicemarks, and copyrights? USPTO, November 16, 2003 [accessed December 12, 2003]. Available from http://www.uspto.gov/web/offices/pac/doc/general/whatis.htm.

U.S. Pharmacopeia (USP). U.S. Pharmacopeia Timeline. U.S. Pharmacopeia. 2003 [accessed July 20, 2003]. Available from http://www.onlinepressroom.net/uspharm/.

U.S. Senate. Senate Committee on Health, Education, Labor and Pensions. Ensuring drug safety: Where do we go from here? March 3, 2005.

U.S. Senate. Senate Labor and Human Resources Committee. Hearing: Final Report of the President's Advisory Committee of the Food and Drug Administration. May 15, 1991.

U.S. v. Sullivan. 1948. U.S. Supreme Court.

van der Laan J-W. New perspectives for alternative approaches to carcinogenicity testing: a regulator viewpoint. Toxicol Lett. 1998; 102–103:561–564.

Vane JR. Inhibition of prostaglandin synthesis as a mechanism of action for aspirin-like drugs. Nature New Biol. 1971; 231:232–235.

———. The mode of action of aspirin and similar compounds. J Allergy Clin Immunol. 1976; 58:691–712.

Vollmer S. Biogen's big lead melting away. Triangle Business Journal. July 19, 2002.

Wang L. Interview with Bernice Schacter. 2003.

Wazana A. Physicians and the pharmaceutical industry: is a gift ever just a gift? JAMA. 2000; 283(3):373–380.

Weggen S, Eritsen JL, Das P, Sagi SA, Wang R, Pietrzik CU, Findlay KA, Smith TE, Murphy MP, Butler T, Kang DE, Marsuiz-Sterling N, Gold TE, Koo EH. A subset of NSAIDs lower amylodgenic Abeta42 independently of cyclooxygenase activity. Nature. 2001; 414:212–216.

White RM. Unraveling the Tuskegee study of untreated syphilis. Arch Intern Med. 2000; 160(5):585.

Willis RC. The discovery doldrums: NCEs are coming slowly; Jurgen Drews explains why. Modern Drug Discovery. 2004; 7(4):23–24.

Witter J. COX-2 & 2x: improved safety? Meeting of the FDA Arthritis Advisory Committee: CLASS/VIGOR 2/7-8/2001.

Wolfe MM, Lichtenstein DR, Singh G. Gastrointestinal toxicity of nonsteroidal anti-inflammatory drugs. N Engl J Med. 1999; 340(24):1888–1899.

World Health Organization (WHO). Guidelines for drug donation. Geneva, Switzerland: World Health Organization; 1999.

World Medical Association (WMA). Ethical Principles for Medical Research Involving Human Subjects. Ferney-Voltaire, France: World Medical Association; 2002a.

Young JH. The long struggle for the 1906 law. FDA Consumer, June 1981 [accessed August 3, 2005]. Available from http:www.cfsan.fda.gov/~lrd/history1.html.

Index

About the Author

BERNICE SCHACTER Ph.D., has over 25 years of biomedical research experience in both academia and industry. She served on the faculty of the School of Medicine of Case Western Reserve University and conducted immunology research at Bristol-Myers Squibb Company. She also served as Vice President of Research at BioTransplant, Inc., a biotechnology startup company in Boston, MA. She has published over 50 papers in peer-reviewed journals and is a co-inventor on four issued patents. Since 1994 she has been a biomedical consultant and writer. She has taught immunology to undergraduate, graduate, and medical students and has developed and offered biotechnology courses for liberal studies students at Wesleyan University in Connecticut and at the University of Delaware.